FEMALES of the SPECIES

Sex and Survival in the Animal Kingdom

Harvard University Press • Cambridge, Massachusetts • London, England • 1986

Printed in the United States of America
10 9 8 7 6 5 4 3 2 1

This book is printed on acid-free paper, and its binding materials have been chosen for strength and durability.

Library of Congress Cataloging-in-Publication Data

Kevles, Bettyann.
Females of the species.

Bibliography: p.
Includes index.
1. Females. 2. Sexual behavior in animals.
3. Reproduction. I. Title.
QL761.K48 1986 591.56 85-27046
ISBN 0-674-29865-9 (alk. paper)

Designed by Gwen Frankfeldt

For Dan, in all his roles

Preface

BEFORE the mid-twentieth century, the female as an active participant in evolution was largely overlooked. Most scientists interested in animal behavior were men who sometimes displayed bias toward, but more often obliviousness to, the often subtle patterns in the lives of female animals. Prominent naturalists in the past wrote as though the oceans were stocked with male fish only, or the skies filled with ganders but no geese. In the last few decades, however, female animals have been recognized as more than passive props in the drama of evolution. Indeed, their behavior has become the preoccupation of a significant number of evolutionary biologists.

These scientists are using new theoretical as well as technological approaches in the investigation of animal behavior. Some venture into the rain forests and over the tundra, where, unlike most of their predecessors, they remain for long periods of time, blending into the environment so that the animals become habituated to human beings and thus allow their lives to be accurately reported. Others complement this field work with captive colonies in zoos, aquaria, and laboratories that provide control populations whose ancestry is known and whose lives can be completely monitored.

Probably no large land animals are left to be discovered, but untold numbers of smaller species, from insects to fish, have yet to be classified. Of the known species, only a fraction have been studied. Yet these investigations have already produced a rich lode of data, revealing both remarkable diversity in the behavior of nonhuman females, as well as patterns that repeat themselves within species and between species taxonomically remote from one another. Despite the great physical and genetic differences dividing invertebrates and vertebrates, it is clear that females across the phyla frequently respond in similar ways to similar challenges.

My aim in this book is to provide the general reader with an integrated report of these new data and the patterns they suggest, interpreted where possible within an evolutionary framework. Of course females do not constitute the whole evolutionary story, any more than males, but what has been learned about the lives of females in recent years strikes me as intrinsically so fascinating—and scientifically so important—as to merit its own account. Thus, I describe here the common features of the female animal's life: the ways that females court then mate with males, nurture their young, and cooperate and compete with one another for survival.

This book began as a joint project with Susan Grether, who guided me into the zoological kingdom which she knows so well. She helped with research and technical information and, both before and after she left the project because of other obligations, remained a reliable fount of wit, continuously optimistic, encouraging, and loyal. I am immensely grateful to her.

Many people gave generously of their expertise to help me complete this book. While I was in Cambridge, Massachusetts, during 1981–82, Edward O. Wilson, Sarah Hrdy, Barbara Smuts, David Crews, David Policansky, Karel Liem, George Tooby, Irven DeVore, and members of his "simian seminars" provided stimulating discussions as well as indispensable leads to work-in-progress that might provide me with more information about female behavior. In England, Tim Clutton-Brock, Brian Bertram, and Robert Martin were very helpful, as were Maggie Redshaw and Alison Jolly. These people do not fit into any common ideological slot, nor do they necessarily agree with one another. Certainly none of them is responsible for the way I have handled the data they introduced me to or the theoretical positions I support.

Early drafts of the manuscript were ably criticized by Garland Allen, Alison Richard, John Allman, Evelynne McGuinness, and Mark Konishi, all of whom offered concrete suggestions. Specialized sections of the book have been scrutinized by Jared Diamond, Mark Konishi, Sandra Verhencamp, David Crews, Biruté Galdikas, Graeme Lowe, Catherine Cox, Paul

Sherman, and Richard W. Burkhardt. I want to thank Carol Verburg and Sara Neustadtl for their literary suggestions, and Peggy Burlet and Maria Abate for their assistance in checking the manuscript for accuracy. The library at the Museum of Comparative Zoology at Harvard University was a cornucopia of information, and the staff there was very helpful. I am also grateful to the staff of the Millikan Library at the California Institute of Technology, particularly Jeanne Tatro in Interlibrary Loans, and to Anne Cain at the Pasadena Public Library.

I am deeply grateful to Susan Wallace at Harvard University Press for her clear thinking, firm guidance, and enthusiasm. To my editor, Howard Boyer, special thanks for his sustained support and insightful judgment. The skills of Amy Malina in tracking down photographs, and of Laszlo Meszoly, whose artistic renditions so enhance the text, were essential to the completion of this book. And finally, my thanks to Beth and Jonathan Kevles for their good-natured support, and to Daniel Kevles, whose help at every stage in every way was immeasurable.

Pasadena, California
September 1985

Contents

part i

COURTSHIP

one

The Female Chooses

THE WORD "courtship" has an old-fashioned ring, calling to mind ardent suitors serenading coquettish virgins. Indeed, until recently zoologists understood all courtship in the animal kingdom to mean the wooing of a female by a male. In this scenario, males were overwhelmed by waves of lust but could not simply beckon or otherwise signal a compliant female to do their bidding. Instead, the male had to win her, cajole, even trick her into behavior that she inherently abhorred. Ambivalent at best, the courted female might tease her suitors and inflame their zeal. Sometimes, it seemed, she would even blackmail them into providing for her material needs before submitting reluctantly to an act that led inevitably to insemination.

These attitudes toward courtship among men and women suffused the views of animal behavior advanced by Charles Darwin, the great progenitor of evolutionary biology. "The female is less eager than the male," he wrote in 1871, "She is coy." And when she does take part in selecting a mate, she chooses "not the male which is most attractive to her, but the one which is the least distasteful."

This view of females reflected in part Darwin's own Victorian moment in history. Brilliant and imaginative though he was, Darwin was rooted in a narrow social tier of a newly industrialized nation that happened to be at the helm of an empire. Not surprisingly, he generalized about the behavior of all human females from his familiarity with the behavior of upper-middle-class English women. This social group was isolated from productive work and discouraged from intellectual pursuits on the grounds that there was a negative correlation between using their minds and bearing healthy children. (Darwin did not seem to be aware of working-class women toiling in mines and factories.) Yet he was a keen and honest recorder of nature, and he conceded "that the female, though comparatively passive, generally exerts some choice."

Observations of sexual behavior in animals led Darwin to expand upon the descriptions of the eighteenth-century Scottish physician John Hunter. Hunter had identified two kinds of sexual traits: the primary, which are apparent at birth, and the secondary, which appear at sexual maturity. Darwin redefined these terms, calling "primary" those traits involved in the act of reproduction, and "secondary" those traits used to attract mates. Implicit in this distinction is the notion that males develop special features in order to attract females, and that females, in turn, actively choose males according to which traits they find most pleasing.

The picture of females as passive, reluctant creatures contradicted Darwin's observations of females as active selectors. Seeking to explain this paradox of female behavior, Darwin looked at the world around him. Little girls, he noted, are as alert and active as their brothers; indeed, in many animal species the young are sexually indistinguishable until maturity. Puberty, of course, brings an abrupt change in mood and demeanor in human as well as animal females. In humans, Darwin argued, their mental progress slows down; and in place of intelligence, females in most complex species develop aesthetic sensibilities, which they use to select their mates.

Darwin called this *sexual selection,* a process he considered coequal with *natural selection* as a mechanism of evolution. Both processes were described in *On the Origin of Species,* which he published in 1859, and further developed in *The Descent of Man* (1871) and *The Expression of the Emotions in Man and Animals* (1872).

Natural selection can be compared with artificial selection, the way breeders produce pigeons and sheepdogs by allowing reproduction among only those with the colors or stamina the breeders desire. Darwin theorized that the natural environment fosters a similar selective process. Survival in nature favors species whose individual members can best find food, protect themselves from predators, and leave offspring. These speculations were supported by Darwin's own observations of animals (especially those he had made as a young man when he circumnavigated the globe aboard the *H.M.S. Beagle*), as well as by those of other naturalists. An extraordinary knowledge of an enormous number of species led Darwin to conclude that only those animals most fit for their environment manage to survive.

Sexual selection takes two forms. The first entails those struggles among males within a species for sexual access to females. As evidence of this male–male competition, Darwin pointed to the special traits, such as sharp antlers or horns, possessed by males in some species that have no other purpose than to intimidate rivals. These appear at sexual maturity as secondary sexual characteristics. Moreover, in many species, especially

Male elephant seals struggle for access to females on Anno Nuevo Island.

mammals, males develop larger body size and greater strength than females of their species—traits that allow them to overpower their competition in the struggle for reproductive success.

Complementing the idea of *male–male competition* for access to females is the idea of *female choice*—the notion that females exercise control over which males will father their offspring. Although some females merely stand back and watch the males of their species fight and then accept the victor, Darwin observed that in many species females select their mates for traits other than strength or aggressiveness. They choose appealing physical attributes, such as the male mandrill's crimson and royal-blue facial markings or the peacock's elegant plumage—traits with no other function than to appeal to the female. However, on occasion the traits an animal needs to survive in the face of a hostile environment are not the

A female (on left) and male blue-winged teal demonstrate sexual dimorphism.

ones needed to attract and win sexual partners. In some species these objectives even conflict with one another. For example, a bird of paradise's voluminous and brilliant tailfeathers, which attract females, can also attract predators and impede the cock's flight. In other words, sometimes sexual selection seems to work against natural selection.

The phenomenon of *sexual dimorphism*—physical differences in size, coloration, and odd appendages between females and males of the same species—was the subject of extensive correspondence between Darwin and his co-discoverer of evolution through natural selection, Alfred Russel Wallace. Each man tried to convince the other that in birds especially, but in some insects and fish as well, the males are brilliantly colored and the females dull for a specific reason. Wallace argued that originally birds of both sexes were bright, but eventually females that were dull and therefore blended into the background were selected. In other words, according to Wallace, females developed camouflage because of its advantages for survival.

Darwin, however, interpreted female dullness as the norm and explained the male's dramatic coloration as being the result of female choice. According to his theory, each member of a species is a unique combination of ancestral traits, and not every combination will produce an equally successful individual. Competition to mate is keen, so that any characteristic which offers an individual some reproductive advantage, however slight, is likely to be passed along. Because females determine to a great extent which males will become fathers and which will not, and because their aesthetic sense often leads them to choose the more beautiful male, over many generations traits such as beautiful feathers in the male become exaggerated. Darwin offered as evidence of this theory, as opposed to Wallace's, the observations that males are more colorful even in those species where males guard the eggs and should therefore be in as much need of protective, dull coloration as the females. He also noted that on islands where similar species do not compete for limited resources, the males are less dramatically different from each other than males on the mainland.

Wallace eventually dismissed female choice altogether as a factor in evolution by defining natural selection so that it included sexual selection. But Darwin stuck to his perception that female choice is a separate evolutionary mechanism. Generally acclaimed by his scientific contemporaries for the body of his work, Darwin found little support for his theory of sexual selection. On the day before his death in 1882, adamant to the end, Darwin would repeat this theory in a notice to the Zoological Society in London. Though he modified his words to emphasize that female choice was not necessarily conscious, Darwin nevertheless insisted that females play a key role in selection and, inherently, in all of evolution.

There were, however, limits to Darwin's vision of the female's role in evolution. Having noted in *The Descent of Man* that the young of both sexes resemble the adult female in most species, Darwin reasoned that males are more evolutionarily advanced than females, since at maturity they continue to develop and change their appearance, whereas females do not. In other words, Darwin did not carry his argument about selection through competition across the sexual frontier. He did not suggest that females, too, might have developed teeth or horns or other traits of their own for the purpose of disadvantaging other females in the struggle to produce more offspring—in short, that females might compete with each other for access to mates. This omission was noted at the time by the American feminist Antoinette Brown Blackwell, a sister-in-law of the suffragist Lucy Stone, in the 1875 publication of *The Sexes throughout Nature.* In reproving Darwin for his oversight, she excused him by assuming that he was simply preoccupied with the enormous "task before him of settling the Origin of all Species and the Descent of Man." But Blackwell went on to note that females must also have evolved through the same process of struggle and competition that influenced the evolution of males. Her criticism, however, was ignored by Darwin and the rest of the male scientific community.

Still-Passive Females

Despite opposition from clergy and from some other zoologists, the theory of evolution by natural selection was accepted rapidly in scientific circles. At the same time, the idea that females contribute to evolution through sexual selection was discounted by all but a few of Darwin's scientific contemporaries. Among the supporters of this theory was George John Romanes, a younger evolutionist and physiologist. Shortly before his death, Darwin handed over to Romanes a great deal of data he had not had time to sort out. Romanes took the data and used it to reinterpret sexual selection. Whereas Darwin had maintained that sexual selection was found in all but the simplest animal forms, Romanes believed it was limited to species that had appeared relatively late in evolutionary time, those higher species that have a sense of beauty. This sense evolved, according to Romanes, as the sexes moved toward more divergent roles, that is, as females became increasingly less cerebral and more emotional. Romanes attributed a superior sense of "beauty" to female animals, as Darwin had, and shared Darwin's view that females were less highly evolved than males—ideas which he articulated in several books and many articles that influenced a generation of biologists.

Romanes apparently saw himself as the guardian of evolution, vested with a responsibility to keep it on the right path. In this role, he interpreted animal behavior in terms of his human preferences, which were to disallow human females participation in the masculine realms of society. Like Karl Marx—who, having discovered the direction of "history," felt compelled to work on its behalf—Romanes argued that evolution, now identified, had to be directed to ensure that human folly did not interfere with the natural order of things.

This attitude was not limited to Britain. At the University of Pennsylvania, the influential American paleontologist Edward Drinker Cope wrote that male animals play a "more active part in the struggle for existence," and that all females, as mothers, have had to sacrifice intellectual growth for emotional strength. He argued that female animals have evolved a strong sympathetic strain, a gift at odds with the masculine skills of unbiased rational behavior. Both Romanes and Cope were describing nonhuman animals but included human beings in their generalizations in ways that reflected the particular human societies they found comfortable.

In the generation after Darwin, biologists began attempting to test some of the hypotheses of evolutionary theory in plant and animal breeding experiments. Spurred by an interest in the evolutionary process, laboratory workers also investigated embryos and cells for insights into the mechanism whereby traits originated and could be passed on to ensuing generations. Darwin, when confronted with this problem, had embraced in part what we now understand as the Lamarckian fallacy (named after its first spokesman, Jean Baptiste Lamarck). This is the argument that traits acquired by an individual animal during its lifetime can be passed on to its offspring. Thus, the giraffe's neck was explained by the suggestion that it elongated from one generation to the next as each generation of giraffes stretched to reach the tastiest leaves at the tops of bushes.

Lamarck's theory was dealt a strong blow by August Freidrich Leopold Weismann, a German cell biologist. In 1869 Weismann examined the reproductive cells of a *hydrozoa* (a small jellyfish-like animal) under the microscope. Watching the "germ cells" divide and the daughter cells repeat the lives of the parents, he was convinced that the germ plasm is immortal. Weismann continued this line of investigation and argued convincingly that, whatever happened later in life, an organism's fate was settled at the instant of fertilization. Nothing that happened afterward could be passed on to succeeding generations, because the substance of inheritance is passed on only at the beginning of life.

This discovery encouraged biologists to focus their microscopes on the egg and the sperm. In 1889 the Scottish biologist Patrick Geddes, with his student J. Arthur Thomson, published the enormously influential book *The Evolution of Sex.* With an introduction by the British sexologist Havelock Ellis, the book argued that the basic differences between female and male begin at the level of the germ cell. The authors characterized as "female" the qualities of passiveness, vegetativeness, and largeness, and as "male" the qualities of high temperature, smallness, and activity. This division translated into two types of physical activity—"anabolism," or constructive metabolism, which was female, and "catabolism," which was destructive and male. These, they reasoned, originated in the fact that eggs are large and nourishing, while sperm are small and absorbing. Females, like their eggs, are passive and stable, while the sperm, hungry for sustenance, is like the male spirit pursuing its ends.

Geddes tried to pacify his critics by emphasizing that the sexes are equal but complementary to each other. "Man thinks more, woman feels more. He discovers more but remembers less; she is more receptive and less forgetful." In Britain and America at the time, angry groups of women were demanding economic and political equality. In apparent response to such pleas, Geddes and Thomson replied, "What was decreed among the prehistoric protozoa cannot be annulled by Act of Parliament." *The Evolution of Sex* went into many editions, spreading this idea of the polarized nature of sexually determined roles into libraries and laboratories all over the English-speaking world.

In the early twentieth century, advances in other areas of biology took center stage from the Darwinists and evolutionary studies. The idea of sexual selection and female choice, always the most controversial part of Darwin's theory, fell into shadow. But the assumption of female passivity was scarcely questioned at all.

Courtship: An Active Strategy

Those naturalists who had earlier interpreted the female's sexual role as passive and coy were not dissembling. They did observe male balloon flies delivering wrapped parcels of insects to their mates, bribing them with food in what looks like an attempt to distract their attention from the matter at hand. Nor were naturalists wrong in observing that the female of the species sometimes appears reluctant to engage in sexual activity. Females in many species do not accept the first available male. But their failure to do so does not necessarily stem from abhorrence of sexual involvement. Rather, these females may be trying out different suitors, and possibly playing for time. Not yet physiologically ready for insemination, they depend on social interaction with the male to stimulate their own reproductive systems, as well as to measure the first suitor against the next.

In many species, as Darwin noted, there is little to distinguish one sex from the other at the infant, juvenile, or larval stage. But after a while the female acquires the secondary characteristics she needs to function as an adult—she develops whatever is necessary to produce eggs and to get them fertilized. Fertilization is seldom a random procedure, although in some species—for example, oysters that release eggs into the water to be fertilized by haphazardly floating sperm—it is difficult to categorize the process in any other way. Most female animals respond to changes in their body chemistry with a series of behaviors designed to seek out potential mates; and where there are several possibilities, females appear to select from among the candidates. Checking out males often consumes a great deal more time than the step to which it usually, but not inevitably, leads—that is, fertilization. This checking-out period is called *courtship.*

Measuring or comparing courting males is what female choice is all about. Darwin was right to suspect that females in some species are attracted to those male traits or appendages which reduce the male's chances of long-term survival. Recent laboratory experiments, for example, have verified observations in the wild that female three-spined stickleback fish *(Gasterosteus aculeatus)* from Lake Wapato, Washington, prefer to lay their eggs in the nests of red-bellied males. Red-bellied males are relatively rare among these stickleback, probably because their bright color makes them easy prey for trout. In birds and mammals as well as fish, many instances have been found since Darwin's day in which the attribute

The female balloon fly accepts a freshly killed insect from the male she is courting.

that makes the male more attractive, and thus more apt to be chosen as a sexual partner, makes him less likely to survive.

For Darwin, a fit animal was one whose strength and health and perhaps natural camouflage allowed it to survive in its environment. To most biologists today, fitness means success at reproduction. They describe an animal's fitness as its contribution to the gene pool of the next generation, compared with that of other members of the species. These two definitions of fitness are not so much contradictory as overlapping for most species. Female birds, for example, may select their mates because they seem both physically attractive and strong enough to help feed hatchlings.

The last few decades of observations in the wild and in the laboratory leave little doubt that in many species females do choose their mates. In analyzing their criteria for making these choices, however, most biologists have departed from Darwin and stopped allocating aesthetic sensibilities to nonhuman animals, male or female. Instead, students of animal behavior today interpret elaborate feathers or birdsong primarily as signals to inform the female that a male of her species is in the neighborhood and available to mate. They explain some male traits, and the female's response to them, as evolutionary adaptations that may have once had an important practical function. For example, recent research indicates that the gorgeous shapes and colors of the male lyrebird's tail are not what attract females of that species, but rather its length, which increases with age. The female may best be able to judge the maturity of a male—his ability to survive for many seasons (and hence his ability to sire many healthy offspring)—by the length of his tail.

In an enterprising experiment in Kenya, the Swedish ornithologist Malte Anderson studied a species called the widowbird *(Euplectes progne)*, a kind of weaverbird. The adult males are solid black except for one red epaulet atop their wings. But their most conspicuous feature is twelve long tailfeathers that, in flight over the Kinangop plateau, expand vertically into a deep, long keel. In contrast, female widowbirds are mottled brown and short-tailed. Anderson decided to see what it was about the males that attracted females. He selected nine groups of birds, in each of which he color-tagged four males. He cut off the tail of one male in each group and glued the tailfeathers to the tail of another male. The remaining pair in each group served as controls: he left one bird untouched, and cut and reglued its own tail back onto the fourth. Before the "treatment," all the male birds within each group had about equally long tails and equal mating success. But afterwards, the birds with the newly elongated tails fared best, and those with truncated appendages fared substantially worse. This experiment, the only one of its kind so far, shows that the particular females involved responded strongly to the male's long tail, and not to the rest of the bird in front of it. Whether the females recognized the ornament as a marker for overall fitness, or whether they were simply responding to momentary whim, is unknown. But the study does show that when given a range of choices, these females, at least, exercised options.

Long tails or oversized antlers, and their disadvantages to individual survival, appear to be the price males in some species must pay to have a chance at fatherhood. Amotz Zahavi has suggested that these encumbrances illustrate the *handicap principle.* Any male that can survive and propagate *in spite of* unwieldy horns or feathers must be that much more fit in every other way than his competitor, and so by definition is the truly superior beast.

Alternatively, the population geneticist R. A. Fisher concluded in the 1930s that in birds of paradise, for instance, females which select longer-tailed males, and pass along to their daughters an inherited preference for longer-tailed males, thus drive male tail lengths longer and longer until they become an encumbrance. Fisher termed this process *runaway selection,* and predicted that in such instances female choice would lead eventually to the decline of a species. Paul Harvey and Steven Arnold have re-examined Fisher's data and supported his conclusions with newer work. In their view, female choice could result in such an acute decline in male viability that the species would become extinct. Still other researchers have suggested that such an outcome would be contrary to the behavior of living animals. They argue that, statistics notwithstanding, females will mate, and if the male of choice is unavailable because he could not survive, they will settle for second best. This will lead eventually not to the extinction of a species but to a limitation in the secondary sexual traits of the male. Whatever the actual case, these discussions reveal that no self-respecting evolutionary biologist can discuss evolutionary theory today without giving serious attention to the role of female choice.

two

Making Up Her Mind

FASCINATING as it would be to invade the psyche of a sow, a hen, or a mangabey, this is impossible. Some biologists still occasionally trip over their tongues and attribute human ambitions and sensibilities to animals whose minds must remain a mystery and whose behavior we can only know from the outside. It is tempting, but wrong, to attribute careful planning and a judicious weighing of the costs and benefits of one suitor over another to any animal, or to assume that any individual animal possesses foresight. But we can deduce from the way animals behave that evolution has brought about testing strategies whose results give the appearance of careful decision-making.

A courting female has both short-term and long-term goals: she is seeking some kind of immediate physical satisfaction at the same time that she is selecting the male with the best genetic contribution for her offspring. Though the goal of having healthy offspring is obvious to the biologist, and crucial to the species, we cannot assume that the young doe knows she will eventually give birth and need a home, food, and protection for her young. In some species the selection process looks like a mere beauty contest. Yet the doe's response to the stag's elaborate antlers and great size and the ensuing courtship ritual make it clear that she is indeed checking out the potential father for his contribution to the survival chances of her offspring—while simultaneously pleasing herself. It is not surprising that an animal's immediate needs and long-term goals usually agree, because natural selection has cloaked the genes of an evolutionarily suitable mate in an attractive package.

Behavioral biologists believe today that courtship is crucial to the female animal in several ways. It helps her eliminate inappropriate candidates (such as members of similar-looking species with whom she can mate, but who will produce either no offspring at all, or offspring that are themselves sterile) and to select from those that are suitable the one which offers the best chance of producing fit offspring. In so doing she safeguards the continuity of the species and influences its evolutionary direction. In all sorts of animals, from fireflies to lizards, males attempt to fertilize any possible candidate, while females check out and refuse imposters. Females and males in the majority of species have a different approach to reproduction. Males produce far more sperm than females do eggs, and it is in the male's interest to fertilize as many females as he can. But an egg is a greater investment for the female—both in terms of the size of the egg itself and in the time and energy necessary to help it mature into a viable

A female red deer acknowledges her harem master on the Isle of Rhum.

organism—than the sperm is to the male; so the female is more protective of her eggs than the male is of his sperm, and is therefore wary of potential suitors. Through courtship, females in most species can take the time to assess a male's ability to provide them with protection, shelter, and nourishment for themselves and their progeny.

Criteria for Choice

Females in many species judge these capabilities directly. They merely stand back and watch males fight until the victor is ready to mate. The female usually accepts him, perhaps because her own male offspring will benefit from his genetic endowment when they in turn must fight reproductive battles of their own. Females may even initiate conflict among the males, as do the northern elephant seals *(Mirounga angustirostris),* which have been studied up and down the coast of California for more than a decade by the mammalogist Burney Le Boeuf. At sea in the Pacific most of the year, females emerge from the water on the coast of California and Mexico only about seven days before giving birth. They gather in groups called pods, which contain from a few to hundreds of females. Here they nurse their youngsters under the eyes of several enormous bulls. The males place themselves strategically among the crowds of females, waiting until about a month after the births, when the females are ready to be inseminated again. One great male dominates access to a pod throughout the peak of the season, but as time passes, he may wear himself out in terms of both general energy and viable sperm. During copulation the females sometimes utter a kind of snarl which seems to encourage other bulls to intervene. Thus the females incite the males to compete among themselves, as if to ensure that only the very strongest male will actually father next year's pup. Just as the females are ready to return to the sea, they change tactics, by courting the younger, smaller bulls which, up until this time, have been unsuccessful bystanders. Le Boeuf hypothesizes that this strategy ensures the female fresh, viable sperm so that, one way or another, each female will leave the island with a fertilized ovum.

In elephant seals as well as other species, females find a new breeding male every year. The male's great size enables him to monopolize many of the smaller females; yet his career, like that of most human athletes, is a short one. The male who must defend his female "property" against competitors, and at the same time try to impregnate those females, exhausts himself within a season or two.

Distant from these great beasts on the branches of the evolutionary bush is the brilliantly colored male jewelfish *(Hemichromis bimaculatus).* Like other freshwater cichlids, they engage in violent and athletic maneuvers in their tropical African ponds. Scarcely four inches long, jewelfish dart to attack, nipping and pushing enemies many times their size. The male is an aggressive suitor upon whose bellicosity the female can eventually rely when there are babies to defend. Thus some females choose males for the talents the males promise as fathers, or fathers-to-be.

The ornithologist Marion Petrie's study of English moorhens *(Gallinula chloropus)* showed that the females choose—and even fight among themselves for—the shortest males who happen to be the fattest. Those pairs with the short, fat males started more clutches earlier in the season; and

Female moorhens bicker over possession of the smaller, fatter male.

since males in this species do most of the incubation, the fat male's large reserves enable him to incubate the eggs longer. Thus the fattest males are the best fathers.

Other females choose males for the shelter they can provide their incipient offspring. Animals cannot, of course, find a "customized house," yet females of many species seem to expect one, and males have evolved, through selection, the skills to supply them. Among fish, the bubblenest builders, small Anabantidae, live in freshwater ponds in southern Asia and Africa, where they feed upon tiny water animals. Among the best known are Siamese fighting fish *(Betta splendens)* and paradise fish *(Macropodus operculatis).* Unique to this family is the labyrinth, a special breathing apparatus located in the head that helps extract oxygen from the air. To use it, these small fish have to swim to the surface and gulp air through their mouths. When it is time to breed, the males suck in an air bubble, but instead of using it to breathe, they cover it with saliva and release it to float on the surface of the pond. Hundreds of bubbles later, the paradise fish has completed a many-chambered nest which he floats to land, often under a mound of leaves. After courting the female and literally squeezing the eggs out of her, he fertilizes them and encloses them inside his bubblenest. Once they are there, he redoubles his bubble production as the old bubbles gradually burst and he must produce new ones to support the eggs. Although the male courts the female, she chooses the male for his nest.

Choosing a male by his nest is a common way of picking a mate among birds. The males usually construct bowl-shaped vessels out of grass, twigs, their own feathers, and, depending on the species, almost any other kind of object. The nest has to be strong enough to hold both eggs and adult birds. Some male birds build splendid edifices which they offer as apparent nuptial prizes. A male wren, for instance, will build four or five nests throughout the year. As they decay, he moves on, never repairing those he has used before; when courting, he leads a female to his freshest creation. If she accepts it, she accepts him and puts the final touch, the lining, into the nest herself. (The abandoned nests are sought after by young birds newly expelled by their parents, who huddle in groups up to a dozen in number in what have been called "dormitory nests.")

All communal nests are not the discarded homesteads of these small wrens. Sparrowsize black village weaverbirds *(Ploceus cucullatus)* build large "apartment complexes" all over southern Africa. They look like small, thatched cottages on the limbs of a variety of trees. The weaverbirds are group-living animals, and more than two hundred individuals may build their nests in a single tree. The male selects plant fibers which he braids into a ring perpendicular to the branch. He then weaves this ring into one giant room with smaller antechambers leading off it. No simple bowl, the weaver's nest is roofed and insulated from tropical rainstorms. The male weaves steadily all day long, never even raising his wings. Only after he has completed his addition to the communal nest does he court a female. Dangling head-down from his new home, he beats his wings, revealing bright yellow underwing feathers, whenever a female approaches. He announces his availability in a high-pitched caw that sounds like "look see, look see." If a passing female is attracted, she will inspect his building. If pleased, she will add a bit of lining grass and, in effect, move in. Females are a precious commodity among weaverbirds, and the competition for them is keen. If several females reject a male's nest, he will usually abandon what he has spent the day weaving and start anew in the morning.

Females sometimes select males for the food they promise to provide. Whether sitting on her eggs or traveling about with a growing infant inside, many females depend on a male to provide them with sufficient nourishment. When courting, a male may offer a tidbit as "proof" that he will be a good provider in the future.

In the deserts of California, Arizona, and New Mexico, the male roadrunner catches a mouse or baby rat, then holding it in his mouth, raps it to death or into a state of shock by pounding it on the ground. He shows it to a female, who may beg for it. But the male withholds the food, all the while waving his tale and croaking. The two birds copulate, with the male still holding the mouse away from the female. Only afterward does he release the gift, which will provide nourishment for the eggs he has just fertilized.

A few females choose a male for the territory he defends. Small female African bushbabies move into territories apart from where they have grown up, apparently selecting forest with a suitable food supply. That territory is usually occupied by a male bushbaby, and the female accepts the male with the territory. There is an assumption here that any male strong enough to dominate a territory is more fit than other local males of the species.

But in some species females select males which contribute nothing to the next generation except their genetic endowment. In these species, females and males live separately from each other most of the time, meeting only for a short period of courtship in a special place called a *lek* (a Swedish word meaning "to play"). The peculiar nature of leks has attracted a great deal of attention because lek breeding has evolved independently in so many widely varying species—birds especially but some mammals as well as fish, amphibians, and insects. All lekking animals choose the same display site in which to gather year after year. The males arrive in mass and herald their arrival by either making a great noise, if they are bullfrogs, or flashing lights, if they are fireflies, or in some other way alerting the scattered females of their species that they have assembled. As if responding to an advertisement, the females hurry to the lek, where they select from among the many displaying males the individuals with whom they will copulate. Once mated, females in lek species depart to bear and raise their young by themselves.

In his important study of the sage grouse *(Centrocerus urophasianus)* in eastern California, R. H. Wiley noted that during the lekking season the females simply stand by, watching what looks to outsiders like a beauty contest. The males fight over a piece of land which, once won, they use as a stage for performing a complicated dance. Pirouetting, they show off the thick ruff of white feathers around their necks, their colorful underwing feathers, and their pointed tails.

The degree and criteria of female choice in leks is still debated. Early evidence that many females select the same male seemed to indicate that the chosen male had the best territory, and that his elaborate performance was gratuitous because the females were really opting for the territory, not the male. But new research by Jack Bradbury seems to disprove this hypothesis. It appears that females in some bird species choose the most vigorous and longest displaying male. Whether it is for the territory he has managed to hold, or for his other qualities, one or two males do indeed attract the preponderance of females. Almost every female and the other males clump around these favorites. However, since males in lek species do not contribute anything to the survival of the next generation beyond their sperm, it is not clear on what grounds females make these choices, if in fact they do make choices.

That they do somehow choose is suggested by evidence among sage grouse, and also among hammer-headed bats (another lekking species), that females move from lek to lek. Thus, even if they allow the males to battle it out in one lek and inevitably accept the winner of these intramale struggles, the females exercise some choice by selecting which lek they will visit. As for the females choosing the territory and not the male, among sage grouse this has been disproved by evidence that females lose interest in a territory when certain males there are supplanted, and seek elsewhere for mates.

A male sage grouse displays his plumage at the center of a lek.

Lek behavior is an intriguing phenomenon because it isolates the genetic benefit of a male from whatever other talents he may offer in providing nourishment, defense, and parental care. In some instances, no doubt, males settle between themselves which male shall mate. But in other cases females do the choosing, though their criteria for selection remain a mystery.

When Nothing Else Will Do

Protection, shelter, nourishment—these needs are surely crucial to most females. But they may be less urgent to some than immediate sensual gratification. It seems apparent that the sperm-laden male of the species is eager to deposit his product inside the female or, in the case of external fertilization, in close proximity to her eggs. Female animals apparently have similar drives. Although some females enter courtship needing to build up their readiness to mate (and the same can be said of the males of these species), it is also apparent that other females seek males for immediate sexual gratification.

Contrary to the prejudices of nineteenth-century zoologists, a glimpse into the animal world is enough to convince anyone that fertile females try to mate with their male counterparts. Indeed, females often initiate the complicated rituals that eventually lead to sexual union. There is little doubt that most females seek sexual liaisons and that they do not need to be convinced, coerced, or even tricked into sexual activity, though they do have to agree to it.

This often occurs at the point of ovulation, when the female's hormones send her in search of a companion. What is not so easily explained is the motivation of many female animals, particularly birds and mammals, and even more particularly primates, which seek sexual contact when they are either already pregnant—so that the act is quite literally wasteful—or when they are too immature or simply at the wrong stage of ovulation to conceive.

In 1969 the ornithologist Larry Wolf studied the purple-throated carib hummingbirds *(Eulampis jugularis)* in the cacao-banana plantations of Dominca in the West Indies. He discovered that in the winter months before the females can produce eggs, the males dominate the nectar-rich fields of flowers. The females approach the males, hovering near them and soliciting copulation. While preparing for this act, the female takes her fill of nectar, returning many times to engage in sexual activity although she cannot possibly lay eggs. This activity can be viewed as "prostitution"—indeed that is what Wolf suggested—a trade-off of sex for food. An analogous situation has been reported by Suehisa Kuroda among the pygmy chimpanzees *(Pan paniscus)* in central Zaire. These small chimps are unique among primates in their food-sharing habits, distributing not only the rare kill that provides them with meat but sharing abundant vegetable foods such as sugar cane and fruit. Females frequently solicit food from males, and Kuroda observes "that females are more successful in getting food from males if they copulate first." It is probable that the females who offer sex in exchange for nourishment are hungry for both food and the sexual activity that they perform to obtain it.

Pregnant females seeking sexual activity have been observed in several mammal species; this is called *pseudosexual behavior* because it cannot result in offspring. Young African elephants *(Loxodonta africans)* living in

Central members of a group of pygmy chimpanzees socializing in Zaire.

female communities have been observed seeking a sexual partner halfway through a 22-month pregnancy. A number of primates, including baboons *(Papio)* and chimpanzees *(Pan troglodites),* have also been seen soliciting courtship although already pregnant. In a carefully observed group of captive rhesus macaques *(Macaca mulatta),* females in the first third of their pregnancies seemed especially interested in winning the sexual attentions of males in their community. And the obliging males, in these instances, responded to the active solicitations.

The entomologist William Eberhard argues that copulation for many species is itself part of the courtship process. Males in these species have penile appendages whose only function appears to be to provide the female with tactile stimulation. For example, the spider *(Lepthyphantes leprosus)* inserts empty pedipalps into the female a number of times before he fills them with sperm from a primary genital opening in his abdomen and then transfers those sperm to the female with whom he has been "copulating." Likewise, Eberhard points to structures on the male genitalia of some molluscs as well as worms whose sole function is to give tactile stimulation to the female's genital area.

Clearly, this kind of activity represents courtship rather than the next step in the reproductive process—fertilization. Courtship, whatever its evolutionary origins and purposes, has in some species become an end in itself. And sexual indulgence and the mutual sharing which is part of the process have become desirable as such and thus new items in the behavioral equation.

The evidence suggests that behavior which was once considered unique to human beings—perpetual sexual receptivity—is found in females of other species. And those who once argued that this constant availability encourages pair-bonding or monogamy face evidence now that some female animals may use sexual attraction to encourage polygamy—having more than one partner—or promiscuity as they indulge their sexual appetites to play one male off against another. Once courtship behavior involving sexual activity evolved into a goal in itself in some species, and became separate from fertilization, sexuality may have emerged as a device to enhance long-term relationships between individual animals in species where the offspring need two parents to protect and provision them. By maintaining the male's fidelity, the mother ensures permanent cooperative support for her young. Sexuality can also be seen as a strategy females use to prevent other females of their species from having sexual access to males.

Female mammals possess a clitoris, an organ whose only function seems to be to give sensations. And courtship among some primates that have been closely scrutinized—the pygmy chimpanzee, for instance—includes a generous amount of foreplay. What apparently inclines female primates to such leisurely courtship behavior is the possibility, perhaps the expectation, of orgasm. Of course it is not possible to quiz these individuals as to the nature of the experience, but observations, both in the laboratory and in the wild, indicate that these females do experience something akin to the tension-releasing response of human females. Whether these female animals experience the same kind of pleasure as the human female is not really important. What is notable is that the possible expectation of something special prompts her to seek a mate. A female fortunate enough to have a choice of males may select the candidate who can provide her with immediate gratification. He may also be the candidate best able to protect both herself and her offspring from hunger, exposure, and predation.

three

The Rites of Courtship

HOWEVER the choices are made that bring female and male together, the process of separating themselves as a pair, apart from the rest of their species, is often just a beginning. As a couple, they may next proceed with a complicated ritual before consummating their partnership. These rituals fascinated Darwin, who pointed out that social communication probably evolved from gestures which had a quite separate function in the evolutionary past. A groundbreaking exploration of this idea was published by Julian Huxley in 1914, when he analyzed the ritual gestures of the great crested grebes.

Among these seasonally monogamous birds, the males and females look remarkably alike. Courting while swimming, the male raises his short wings and stretches his long neck, exposing gray and black plumage. Head hunched, he paddles toward the female. She nods to him and soon they are wagging their heads in unison. In perfect synchrony, both birds abruptly flip and dive to the bottom of the pond, resurfacing with a shred of weed caught in their bills. Facing each other again, they dangle the weeds, once more in unison. Huxley interpreted this as a vestige of what once might have been mutual feeding. Weeds in beak, the grebes rise high out of the pond and sway together, breast to breast, then dive again, and dance again, until they are ready to go off to build a nest together. The male constructs a special "stage" in addition to the nest and from its height displays his great shaggy crest, nodding until the female joins him. No longer a mirror image of the male, the female grebe flattens her own crest as she drops her head, awaiting insemination. In analyzing the birds'

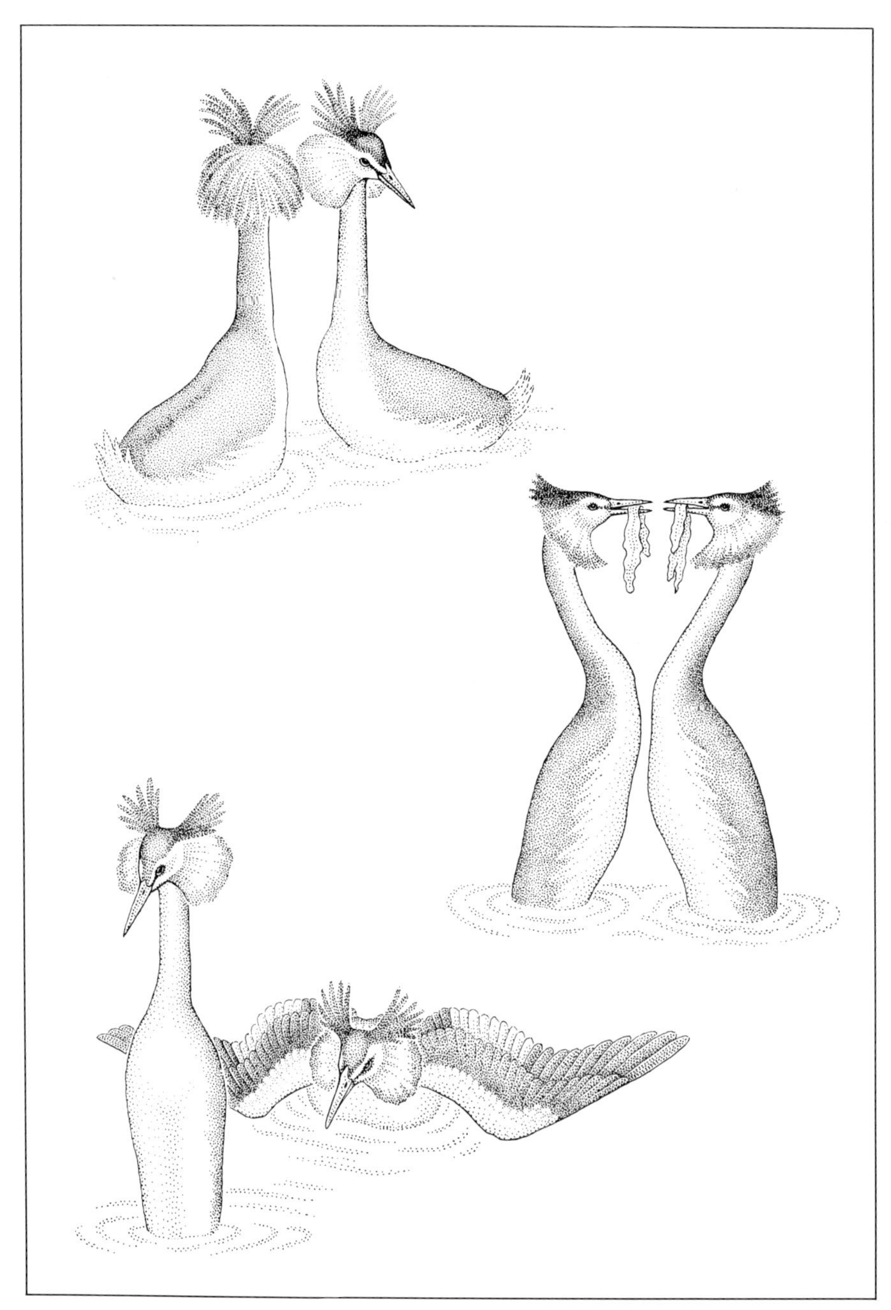

The courtship ritual of the great crested grebe.

stereotyped courting motions, Huxley explained that the display probably began as a simple landing gesture as the grebe approached the nest and evolved to a complicated greeting. He labeled this transmutation of gestures *ritualization.*

Genes, Evolution, and Ritual Behavior

Early in the twentieth century, the study of animal behavior was primarily the domain of psychologists who, especially in the United States, focused on the animal as a kind of machine divorced from its natural environment. The part of Darwinian theory which had proposed that natural selection alters behavior as well as body shape was in a period of eclipse.

Toward midcentury two major advances reshaped the study of the evolution of behavior. The first was the appearance of several important works—starting with the publication of Theodosius Dobzansky's *Genetics and the Origin of Species* in 1937—which brought together into a single theory the modern science of genetics and Darwin's idea of natural selection. This unified theory has been called the *modern synthesis,* and was strengthened by contributions from the population geneticists R. A. Fisher and Sewall Wright and from the evolutionary biologists Ernst Mayr, George Gaylord Simpson, and Julian Huxley. In his book *Evolution: The Modern Synthesis,* published in 1942, Huxley explained that genetic change, in the form of an occasional favorable mutation but more frequently in the form of more gradual, less dramatic recombination, made for organismic change. Further, he pointed out that underlying all evolution was a single basic mechanism: Darwinian selection acting upon the expression of an animal's genetic endowment. Huxley, for one, emphasized that selection ensures the survival and establishment of a long series of "exceedingly rare events"—favorable genetic changes which, with the help of sexual reproduction, are secured into a single strain, or species, of living matter.

The modern synthesis reinvigorated Darwin's theoretical position in almost all its parts, including the distinction between natural and sexual selection. Huxley defined fitness as the ability of an individual to survive to sexual maturity. Fisher and Haldane went a step further by emphasizing the fitness of single genes. Thus they described an animal's success in life in terms of its contribution to the gene pool of the next generation—its ability to leave offspring.

The second advance was Konrad Lorenz and Niko Tinbergen's contribution to ethology, a new approach to animal and human behavior. The word "ethology" was coined in 1859 in France, the year that Darwin's *Origin of Species* appeared in England. Its creator, Isidore Geoffroy Sainte-Hillaire, defined ethology as the study of the characteristics of animals in their natural habitat. Adopted largely by students of marine organisms, which are difficult to remove to laboratories, ethology developed on the continent for its first century.

A rigorous method of observing natural behavior was clearly needed. When the Franco-American adventurer Paul du Chaillu captured a young gorilla in the Congo in 1860, he described the ape's behavior with theatrical embellishments. He was hardly at fault, for there were no rules for him to follow then, no scientific approach or understanding of how to record animal behavior. Ethologists brought discipline to the world of animal watching. The same problems that faced du Chaillu, that of objectively describing and interpreting the behavior of animals in their own habitat, became their focus. This new breed of zoologist sorted out what needed to be recorded, organized the gathering of data, and asked questions that made observers look at their animal subjects in a new way. Thanks to Lorenz and Tinbergen, ethology now combines the naturalist's love of animals as individuals with a rigorous methodolgy that looks especially at animal rituals for past evolutionary patterns.

Darwin had noted that a "ritual" is a series of gestures that reflect an emotional state, and he puzzled over the meaning of these gestures in *The Expression of the Emotions in Man and Animals*. In some ways ethology is a return to Darwin's approach to animals. In 1837 when he was going over some of the bird specimens that he had brought back from his voyage to South America, Darwin recalled that two kinds of mockingbirds seemed to be identical in structure, but different only in behavior. Ornithologists did find a structural difference, but Darwin remained convinced that differences in behavior alone could indicate differences in species. Later students of behavior looked on the rituals animals perform much as anatomists saw earlier forms of an animal family in vestigial organs. Ethologists believe with Darwin that behavior has evolved along with structure, and that the analysis of a ritual can reveal the evolutionary history of a species.

Darwin attributed the origins of rituals to "instincts," and he included a chapter on instinct in the *Origin of Species.* Unfortunately, he assumed that everyone "understood" what was meant when he used the term, and the term "instinct" has been and is still used in many ways that everyone "understands" generally, and most people not at all specifically. Darwin discussed the evolution of instincts and the apparent evolutionary origins of the expression of emotions from the grimaces of monkeys to the human smile. Today ethologists and evolutionary biologists assume that a certain amount of behavior is genetically programmed, that is, both inherited and dependent on experience to trigger the appropriate behaviors. Like Darwin, they agree that a highly bred setter cannot be made into a good sheepdog even with the best training; yet the best sheepdog must still be trained to do its job well.

Ethologists study animals in nature, or in a laboratory where there is as natural a situation as possible. Lorenz was reacting explicitly against psychologists who approached animals as if they were machines with interchangeable parts. He observed that animals seemed to be born with certain "instincts." His experience with a family of baby geese showed that there is a special time in an animal's life when it is ready to learn certain things. He interrupted the goslings' development and substituted himself for their mother as the first moving object they saw after hatching. The result was that they followed him everywhere. He called this early learning *imprinting,* and referred to an animal's readiness to learn a specific object or individual animal as the presence of a built-in "school marm." Today we might prefer a computer metaphor and describe the goslings as "programmed" to learn at a "critical period."

Before the Second World War, Lorenz in Austria and Tinbergen, working with stickleback fish in Holland, dominated the field of ethology, concentrating especially on the complicated gestures between individuals of a species. Later Tinbergen expanded his work by including the habitat in his study of an animal's behavior. He described the rituals and gestures of herring gulls in relation to the pressures of their present environment as well as their evolutionary past.

After the war, perhaps as a reaction to human excesses, perhaps in an effort to explain them, the field of ethology attracted a growing number of students. Behavior as an evolutionarily identifiable property came into its own in 1947 when the Dutch biologist M. S. C. Adriaanse published in the first article of the first issue of the journal *Behaviour* his discovery of what Lorenz had suggested earlier—that on the basis of a clear difference in behavior alone, what had been judged as a single species of digger wasps were in fact separate species.

Instincts in Conflict

Carefully observing repeated patterns of behavior, ethologists have continued to separate the rituals of courtship gesture by gesture, and to hypothesize evolutionary explanations for the origins of each. Lorenz identified a group of behaviors that he called *displacement activities*—the animal's reaction to a conflict of instincts. An example of conflicting emotions is the simultaneous eagerness of the female for sexual stimulation and her inherent fear of body contact with any other animal, including a male of her own species. To accept fertilization, she needs reassurance and encouragement. Therefore, the male has evolved signals for overcoming the female's fears, and his own fears too. The results are the stereotyped rituals of courtship, whose purpose is apparently the moderation of tensions between the sexes. At the same time that these behaviors enable the individual animals to control their fears, they also act as mechanisms to prevent interbreeding with similar-looking species: one false step in the ritual aborts the action.

Infant-like behavior seems to reduce tensions in many species because most animals are gentle with their own young. Male tree squirrels, like male birds, soothe their female's fears by making "baby mews," the kind of sounds an infant makes. When courted, the female bullfinch on her part behaves like a fledging, her beak open as if appealing for the male to feed her from the food stored in his crop. Many female birds, such as the housesparrow, practice food-begging early in their encounters.

In some species the food exchange has become highly ritualized so that the gesture remains, but not the food. Male Antarctic adelie penguins, for instance, offer a pebble in lieu of food, and this first stone, when accepted by the female, becomes the "cornerstone" in their nest, and sets the courtship ritual in motion. The imperial peacock *(Pavo cristatus)* merely pantomimes feeding as part of his approach to the hen. Familiar to humanity for four thousand years, peafowl were taken from India to Persia and then to Europe and the Americas. They have been domesticated and artificially selected longer than almost any other bird. The remarkable survival of these animals can be attributed, in part, to the eyelike markings on their tailfeathers. In India the eye is a symbol of Krishna, and so wild peacocks are never killed but live alongside humans, where they earn their keep by eating small snakes and cobra and alerting their hosts to approaching tigers. In competition with other males, the peacock seeks a stage for attracting hens. When individuals do pair off, a complex show ensues. The male faces the female, then turns abruptly and makes pecking motions toward the ground. He circles her, still pecking, as if indicating where she can find something to eat. No food exchange takes place, and yet the eating motions seem to be the heart of the whole procedure, as if a promise were being exchanged.

Within closely related insect species, such as the two-thousand-odd members of the dancefly family, a peculiar courtship ritual involving food seems to reveal an evolutionary history. The male fly typically presents a freshly killed bug, but as an apparent diversionary tactic. The gift food holds the female's interest so that she neither escapes nor turns on the male as prey. In related species, male flies offer females just a small share of their prey, while males in other species wrap the proposed gift in a silken web which absorbs considerable energy for the female to unwrap to get to the kill. A final member of the family provides an elaborate silken package but omits the food altogether, a hollow gift indeed. Females in each species accept the gift and make an effort to unwrap it while the male proceeds with copulation. The line of development between these species has the appearance of a continuous pattern, but there is no proof, nor can there be, of any evolutionary connection. The ritual is a successful way of courting and may have evolved independently in each case, rather than evolving one from the other in allied species.

Feeding in connection with courting exists in other insect species. Female scorpion flies feast off a secretion that the male produces by dropping some of his own saliva onto a leaf. These drops harden into a food that the female appears to relish, for she devours it contentedly while he completes fertilization.

Food gifts, or gifts in general, are not exclusive to winged creatures, insects and birds. Among mammals, for instance, the dog fox presents his mate with a mouse for a starter and male lions share their kill with their consort.

Ethologists speculate that some species have eliminated the exchange of goods from this food-bribing ritual, but have retained the oral aspect of the act. Love birds tap their beaks together in their so-called "billing." Dogs and cats nuzzle one another, and anthropoid apes, particularly the orangutan, court by exchanging long, noisy "kisses" on the lips. In evolutionary terms, then, the kiss is hypothesized as the last vestige of what was once an offering of food used, most likely, as a diversionary tactic.

Zoologists hypothesize that many females, like the American tree squirrel (Scuridae), feel torn between two powerful instincts: they want to escape and at the same time they want to greet the male. A common resolution to this conflict is a ritualized chase. The grey squirrels will run after each other for up to an hour. At the beginning of the spring, the female may teasingly engage the male in a fruitless run because she is not ready to be caught. But as the days wear on, the chase becomes more serious. She jumps from branch to branch, the male close in pursuit, letting him gradually close the distance between them. She leads the way and if he falls too far behind, she turns and checks and stops to let him catch up.

Cheetahs enjoy a good run too. The female sometimes leads the cluster of males she has gathered on a chase around her territory, exhausting all but the one with whom she eventually mates. The males may fight each other while chasing her, but she apparently ignores their interactions and chooses independently. Each chase follows the same pattern. About a 100-yard run, then a tackle in which the cats play-fight, play-bite, and jump on top of one another like house kittens.

Birds also indulge in courtship chases, but in many instances the chase has been ritualized into a repeated pattern of aerial acrobatics, as with the elaborate series of spirals of Wilson's phalarope *(Steganopus tricolor)*, a kind of sandpiper which was studied in grasslands near the Mississippi River. During the spring migration the female flies over her territory and selects a likely male swimming in the river. Once he has become her quarry, she circles him, threatening all other females with her neck retracted, her bill directed straight ahead. If this offensive posture fails to ward off rivals, she will make an all-out attack. Concentrating on the male, she expands her neck feathers for his apparent approval, making gruff wooing noises. If she succeeds, he will take wing and she will fly after him accompanied by other unpaired females. Around and about they circle until the female succeeds in winning her male and they settle together on the pond below.

Flamingos perform a ritualized courtship dance.

Other birds clear a piece of ground of debris and then race around in ritualized patterns. Flamingos (Phoenicopteridae) step into the grass as onto a stage where, never altering the pattern, they perform the chase slowly, like a dance, the steps of which lead to copulation.

These ritualized patterns are also common in insects like flies and bees, as well as in great hoofed animals. The male greater kudu moves in slow, measured steps behind the female, his horns laid low as if to assure her that his aims are peaceful and she need not fear.

R. F. Ewer hypothesized that in the evolutionary past, fighting was an integral part of all feline courtships. If this were the case, then some of the remaining gestures in all feline encounters are vestigial gestures of those primeval fights. Cats like the cheetah always raise one leg as if to ward off danger as the courtship draws to a close. And in the final phase, both

Cheetah copulation concludes with the male's biting the female on the neck.

females and males sink their teeth into the necks of their mates. The mutual "lovebite" often draws blood, a reminder, perhaps, of the reason for the initial fear the female and male have of each other.

Elephant seals also wrestle as they court, and the necks of the females are decorated with scars, mementoes of earlier encounters. Among ungulates, like the graceful African impala, the fighting has become tamed into a pattern known as *laufschlag.* After the male has sniffed and savored the sexually appetizing odors in the female's urine, he raises his lips in a disdainful-appearing expression. This curious quasi-smile is known by the German term *flehman.* Apparent in cats and kangaroos as well as ungulates, flehman involves an auxiliary organ that in many animals is specialized for sexual scenting. This accessory organ of Jacobsen, as it is called, is a pouchlike structure that sits right above the palate and opens into the mouth through a narrow duct. When the bull curls back his upper lip, he is not using his mouth to express disdain but is rather opening the duct so that he can use this "nose." Then he walks quickly through the herd with his neck stretched forward, his nose high and his mouth ajar. With his horns directed backward, and his penis erect, he chases the courting female through the 100-odd females in the herd. Catching up with her, he lifts his left leg to her stomach, a gesture interpreted as a vestige of what was probably a front-leg kick. Many species seem to indulge in a sudden flair-up of mutual aggression right before they copulate, a final expression of the mixed fear and attraction the act portends for the courting pair.

Observing the courtship of more than fifty species of araneid spiders, the British researchers N. H. and Barbara Robinson found three patterns and grouped the spiders accordingly. These patterns seem to reveal how the self-interest of both female and male spiders has led to an increasingly complex ritual. In the first pattern, the female waits in her web while the sexually aroused male hazards the journey across the silk to meet her at the hub. As he advances, he slaps his first and second legs together, setting up vibrations across the web which signal to her that he is a mate, not dinner. These vibrations induce her to allow him to meet her and wrap her body in silken threads woven from his own body, as she has woven the web from hers. In this fashion they consummate their union.

A group of eleven species in the same genus *Argiope* follows a second pattern. The female waits at the hub at first, watching the male cross the dangerous prey area of her web. She feels him reach the hub and stop to cut a hole; he drops a short mating thread and moves onto it. Dangling from it, he rubs his rear legs together so that they send a vibratory motion upward. Stimulated by the vibrations, the female moves onto the new thread and when she reaches the small male, she assumes the correct posture to complete their act. This is still a dangerous time for the male, as the female often attempts to brush the male off her body, bite him, and wrap him up to eat. But these males have got a safety line, and if they sense the female's intention, they can swing away and try again.

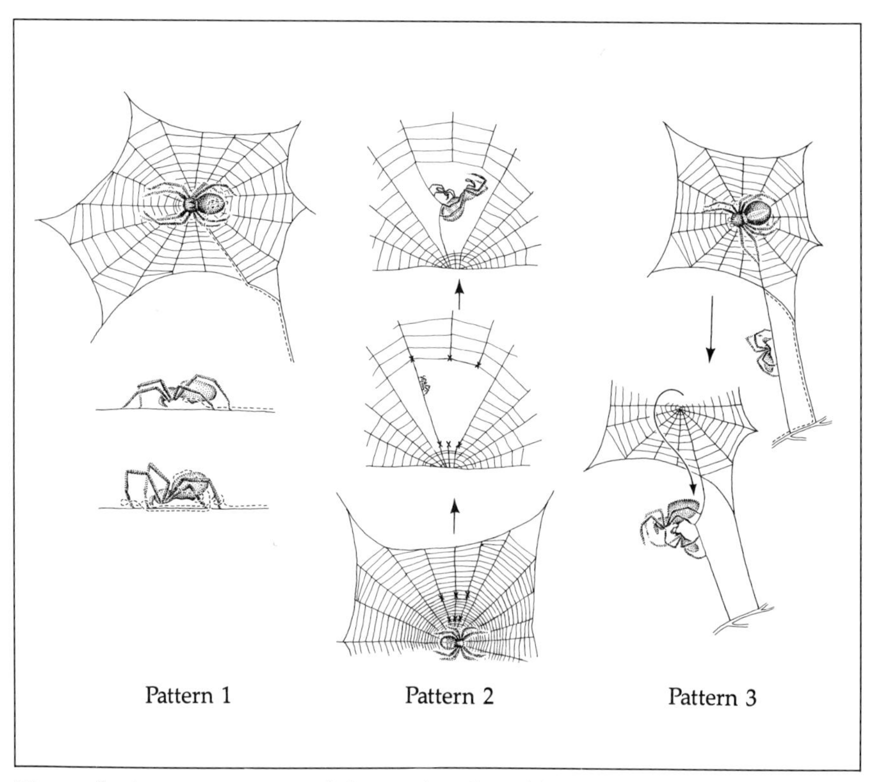

The evolutionary sequence of the mating thread in a group of tropical spiders.

In the third group of spiders, the male spins a mating thread altogether apart from the hub of the female's web. She senses his complex series of vibrations along her own web, and leaves it to join him. The three behaviors represent perhaps three stages in the evolution of a courting system, though we cannot know this for sure. This last form is advantageous to the female as well as the male. She may be deprived of a meal, but now she has the opportunity to measure the courting male by the skill of his courtship, and so enlarges her range of choice. She also has more males to choose from because she is not reducing the supply by inadvertently eating them.

The Robinsons observed female spiders watching two males perform this mating-web maneuver at the same time. The females apparently compared the vibratory motions of each male on the web and selected between them. Within this group, female spiders have evolved from apparently predatory aggressors into compliant cooperators with males of their species. In exchange for giving up the males as possible dinners, females in the last group have a longer, tactile courtship and can be more selective about their mates. The males, while risking refusal, are better off for not having to risk their lives. All of these species of spiders coexist, giving us a rare opportunity to see a pattern of behavior as it may have evolved. But without fossil evidence, it is impossible to know with certainty which of these spiders appeared first. Assuming that the safer process was the last to develop, one might hypothesize that the higher-risk mating procedure may one day spell extinction for that species, but not necessarily. Those spiders unchallenged in their ecological niche may thrive because there is no pressure on them to reproduce differently.

As chases seem to lead to dances, warning screams, another outlet for animals in fear, seem to have been modified into courting signals in many species. The complicated duets of the grass warblers *(Cistiola)* have turned the warning notes to cries of anticipation. Likewise, the female siamang lets her mate know that she is not afraid by singing harmoniously. Yet we lack the living evidence for the evolution of these rituals, the missing behavioral links, that we have for the spider's mating thread. We can analyze a dance, or song, or kiss, and see a history of evasive movements, yet there are no living practitioners of the earlier modes in most species, whether the rites evolved toward complexity or toward simplicity, as there are still spiders spinning simple orbs. The hypothesis explaining these evolved rituals, though forceful, remains speculative. Whatever their origins, these rituals absorb a precise amount of time and seem to play an important role in preparing two individuals of a species to mate. Each gesture the male makes stimulates the next step from the female, which in turn triggers his next move—singing the next note or nodding his head—which she seems to await before proceeding herself. In other words, the drawn-out ritual allows their bodies to arrive at the penultimate condition for the final step at the same time.

The precise relationship of rituals to physiology, and the adaptive advantages in terms of survival and reproduction, have been given varied interpretations. Yet there is no disagreement that these often complex courtship rituals stimulate female and male reproductive hormones and ensure that they fall into synchrony for successful breeding.

For some species this means a long consortship. While a female chimpanzee may appear promiscuous at the start of her sexual season, when she reaches her peak she may suddenly select one favored male and disappear with him "on safari," as some observers call it, to be with him alone for the three days during which she probably conceives.

Even longer is the ritual engaged in by bowerbirds in Australia and New Guinea. Early in autumn, individuals of both sexes live independently. It is not until spring that the female lays her eggs and raises a family. Yet in autumn male bowerbirds stake out arenas, small territories which they clear between food-producing trees. And in these spaces, according to their species, they construct great edifices of twigs and brush, perhaps large "maypoles" about nine feet high, walled avenues or stick towers, all decorated with colored objects like fruit and flowers or human flotsam, if that is available to the birds.

Some bowerbirds employ twigs as brushes, which makes them one of the few animal species that observers have found using tools. While the female bowerbird feeds at random, the male paints his bower and decorates its walls with orchids, perhaps, or pieces of found glass. The magnificent edifices struck the nineteenth-century naturalists who first discovered them with wonder. These bowers helped convince Darwin that female bowerbirds, and by extension all female animals, possess an aesthetic sense. After the male completes his bower, a female will notice it if he is lucky and move in close to watch. For the next three months the male keeps the bower fresh, frequently carrying colorful fruit in his beak for long periods of time. This suggests that perhaps at least part of the ritual has evolved from courtship feeding, as the female is always his audience. The male replaces leaves and flowers daily, and he performs a dance and sings a wide range since he is a vocal mimic. Through it all, the female remains an audience, absorbed enough so she does not go away, but still just an observer, until months later in early spring she begins to ovulate. Only then does the male inseminate her. She then flies off to a separate spot where she builds a simple, undecorated nest and lays her eggs.

The purpose of the elaborate bower seems to be twofold: it keeps the

The male bowerbird appears to compensate for its drabness by constructing an elaborate bower to attract the female.

female's attention until she is sexually prepared to encourage the male to mate, and it enables her to judge the male's vigor, for only a strong male can afford to spend hours a day at the bower, replacing wilted decorations and improving its design as opportunities arise.

The bowerbird's magnificent, elaborate artistry is important in keeping a potential mate in tow, but his work plays no role in the actual courtship process, itself a separate set of ritualized steps that do not take place in the bower. Likewise, the songs of many species of birds are far more complicated than necessary to send a message to a mate. Though the heart of a ritual may have had an evolutionary function, a practical purpose when it first began, rituals may become so removed from the original functions of the component parts that the initial purpose is lost or overwhelmed in the elaborated pattern.

Rituals continue nonetheless, even when the reasons for which they may have evolved are no longer obvious. All behavior, as well as all anatomical structures, probably had some function in the evolutionary past. But just as we retain an appendix which at some distant time may have functioned as part of our digestive system, cats continue biting each other's necks as they copulate. There is not necessarily a purpose for every current structure, or a functional reason for every contemporary gesture. What had a quite different use in the evolutionary past can sometimes linger into the present because it is harmless and simply not relevant to the fitness of the living animal.

four

Sending Signals

THE QUIET silkworm moth *(Bombyx mori)* emerges from her cocoon as night falls in the warm forest, a full complement of mature eggs in her ovaries. Almost immediately she fastens a silken thread to a twig and hangs upside down, her abdomen protruding so that she releases a powerful chemical that volatizes as it hits the air. From this "calling position" her scent travels outward to male silkmoths up to a distance of two miles. Depending on the wind, she may attract a plethora of mating candidates.

The volatilized chemicals are called *pheromones.* They differ from the other chemicals produced by the body because they are intended to influence an external target—in this instance, the male of the moth species. Most sexually targeted pheromones appeal to the olfactory organs. The precise function of these sexually stimulating chemicals is still being explored. Originally researchers believed they were confined to insects, but recent studies reveal that pheromones play a large role in the reproductive behavior of most vertebrates as well, including mammals. A few odors may act alone as powerful aphrodisiacs, as with the silkmoth, but in most instances these chemicals trigger other behaviors needed to complete the ritual of courtship.

A detailed study of the courting behavior of the Mexican garter snake *(Thamnophis melanogaster),* by the herpetologists David Crews and William Garstka, pinpoints the chemical processes at work. Well-fed females develop large liver-to-body weight ratios, and these heavy-livered females produce vitellogenin, which is carried in their circulatory systems to the surface of their skin. There, acting as a pheromone, it attracts males and triggers male courtship behavior. Controlled studies of these snakes showed that the female's readiness always triggered the male's response. And the researchers noted that the vitellogenin that attracts the males goes on to produce yolk for the soon-to-be fertilized eggs.

A New England subspecies of the garter snake *(Thamnophis sintalis parietalis)* winters in underground dens containing as many as 10,000 individuals. As spring warms the earth, the males emerge first, en masse, to camp at the mouths of their dens. Because the females come up singly or in small groups several weeks later, the ratio of males to females can be as great as five thousand to one. Each female attracts as many as one hundred males with her fragrant pheromone, turning the crowd into a writhing ball of squiggling snakes. Those males that reach her first flick their tongues up and down her back, drawing the chemical over the roof of their mouths in order to receive the pungent message through an organ of Jacobsen, much like the African impala's. Although the intensity of the female's pheromone is constant, the male's subsequent rubbing of his chin backward and forward along the female's back may make her breathe more deeply so that she stretches her skin and allows more of the chemical to seep through.

Mammals as well as reptiles and insects use their tongues to search for cues about their partner's sexual state. Most tasting, as opposed to smelling or use of the Jacobsen's organ, focuses on the genitals, but it is difficult to separate urine-tasting, which detects readiness in both males and females, from genital stimulation, the caresses observed in animals as diverse as elephants, lions, and orangutans.

Females may send out and respond to a range of signals, some quite obvious to a human observer, such as an orangutan's deep belly call, and others more subtle, like the undulation of a goldfish's tail. Ethologists emphasize that there is no sensory hierarchy among mating signals; that is, one kind of signal is not more successful than another. Visual signs are not superior to sounds, nor is odor a lower order of attractant than flavor. There is a difference, however, as to the amount of control an animal has over *receiving* a cue or reacting to it. Mammals, for instance, can open and close their eyes so that visual cues can be in a sense "looked for." Mammals cannot close their ears, though, and so sound becomes an involuntarily picked-up signal. But the degree to which a signal is voluntary by sender or receiver depends on circumstances. For example, female New

World cebus monkeys chase males when they are in estrus. The females present themselves for copulation, and they vocalize as well. A male will often ignore these signals until the female's persistence persuades him to cooperate. Which sense the male is finally responding to, the visual or the oral, is hard for the observer to discern.

Many animals, in fact, use several senses simultaneously, or in sequence, when they are ready to begin courting. Sight may be the most obvious one, but that does not mean that special sounds or smells do not play an important role in bringing female and male together.

Sight

Visual cues are used by a whole range of animals—insects, fish, birds, and mammals—to initiate courtships. They may not be the only signals particular females use, but they are especially obvious to the human spectator. For fireflies (Lampyridae) who court at night, light is the only signal, but the female uses it in several ways. Often called lightning bugs, these tiny beetles, both females and males, have special abdominal sacs filled with a luminescent chemical. Whether at home in the warm grasslands of southeast Asia, India, or the eastern United States, fireflies court just after sunset. The hundreds of species of these beetles seem identical to human observers; many appear identical to other fireflies as well, some of which are unable to distinguish a member of their own species from a look-alike predator.

Females of one variety in the Florida grasslands hide in tiny earthen burrows throughout the day but emerge after sunset to perch under broad leaves. For a few minutes they flash a coded signal skyward. If they are really ready to court, they emit a species-specific signal and get an answer from males in the vicinity by return flash. Then the night sky is filled with male fireflies looking for a mate. When the male of her choice zooms in, courtship and consummation are completed quickly. Each species flashes a different courtship pattern, some simple, some complex; the female's signal and the male's response can be identical or form a synchronous pattern.

However, the lives of fireflies are rich in deceit. In an ingenious study in 1966, the entomologist J. E. Lloyd sent out imitative light signals with a flash light and discovered that female fireflies may also emit false signals. That is, some females send out the right signal until they have successfully mated; then they change signals to attract males of a different species which will become nourishment for their developing eggs. The variety of these flashes has prompted researchers to compare firefly communications all over the world. Those in Florida act on their own, but in Borneo the insects swarm, lighting up whole trees in unison, casting light on the thick foliage. Because the adult beetles live for only a few days, they must mate immediately, or not at all.

Luminescence is not limited to insects. Fish, too, have the ability to light up the deep waters round them, as evidenced by the lantern fish *(Myctophum punctatum),* a denizen of the middle oceans. Special appendages on their bodies operate like blinkers, uncovering and recovering the luminescent areas near their eyes. Females and males look very different in this species. Although their behavior has scarcely been explored, the winkings and blinkings seem always to be exchanged between the sexes, perhaps as a preliminary to courtship.

Eyes have been a part of courtship for eons. They play the crucial role in the reproductive life of one of the planet's most ancient inhabitants, *Limulus,* the horseshoe crab. Its form unchanged for millions of years, the horseshoe crab is simple in some ways, but not in its vision. Its complex eyes are a clue to the nature and evolution of sight. Not surprisingly, males of this mostly nocturnal species find their mates by sight, moving to mate with the female with the darkest carapace. In a series of controlled experiments, Robert Barlow has discovered that both male and female visual sensitivity increases at night. But night or day, males *look* for mates. Unlike most other species, they have no chemical or tacticle courting cues but depend on sight alone. Still unknown is the relative reproductive success of darker-shelled females, if there is any difference at all, and the possibility that the absense of a dark shell may act to protect newly moulted females, who are gray and hard to see, from being mated before they are ready for what can be a decade-long embrace.

Most species that depend on vision are diurnal, and in daylight color is an important aspect of any visual "flag." Shallow-water fish announce to each other their sexual readiness by shifts in coloration or intensity that are clear in the sun-drenched sea. Throughout a good part of the year, most female and male pipefish (Syngnathidae) are a dark, drab green all over. Looking something like an armored pipe-cleaner, the foot-long fish, along with their relatives the seahorses, are recognizable by their unique shape and by the marsupial-like pouches on the underbellies of the males. When the female pipefish is ready to breed, her head turns distinctly yellow, and she swims over to the male, wriggling her bright head in all directions so that he cannot misunderstand her condition.

The change in the appearance of a female chimpanzee's rump is not as abrupt as the change in the pipefish's head, but it has the same effect on the males of the species, in that it indicates a readiness to mate. When a female chimpanzee comes into estrus, approximately monthly, she sports a raw and swollen posterior. Among chimps at the Gombe Stream in Tanzania, the younger males in particular discover the female's condition by sight,

At night, male horseshoe crabs move toward a black cement simulation of the female.

and then proceed to sniff her genitals. A male's approach is often all that the female needs to assume the mating posture. She may allow each attending male his turn, at least in the early part of her cycle. Primates other than chimpanzees are well aware of the female's sexual stage without benefit of posterior color. Perhaps odor alone is signal enough, and it is possible that this very visible sign of availablity is no longer needed within the chimpanzee community. But it does occur monthly among sexually mature females, which seems to indicate either that at one time in their evolutionary history this visual signal was very important, or that it is merely a neutral trait that was never selected out.

Jumping from fireflies to chimpanzees is a great leap. Fireflies seem able to succeed with visual signals alone, while more complex animals like primates use visual signs to attract potential mates, but then resort to body contact signals, those that rely on smell or touch or taste, to complete the checking-out process of courtship.

Sound

An animal's habitat plays an enormous role in determining which signal or set of signals will work best. Many inhabitants of the rainforest do not rely on bright signs to alert possible mates, because the competing colors and patterns of flowers and leaves in the dense foliage would overwhelm them. Instead, many forest-dwelling females depend on sound, that other remote signal, to alert males of their condition.

Mosquitoes, tiny wisps of noisy action, are almost synonymous with the buzz they make. What most of us seldom notice is the difference in pitch between the female's and male's whine. The buzz comes from the movement of the wings in motion. And the sound of the female's wings is what attracts the male yellow-fever mosquito *(Aedes)* to a mate. Experiments with yellow-fever mosquitoes revealed that a sitting female surrounded by males did not induce a single one of them to approach her. But that same female, flying, attracted hordes of suitors. The researcher presented a freshly killed female to some males, who ignored the corpse entirely. Then to prove the point, the researcher tempted a group of male mosquitoes with sound from a tuning fork of the same pitch as the buzz of the flying females, and the males all rose and zoomed in anticipation. Female mosquitoes ignored the same sound. In mosquitoes, it appears, the female sends the signal and awaits suitors but does not respond to the male's wing-noises, which can be as much as an octave higher than her own.

In a few species of birds, such as boubou shrikes *(Laniarius aethiopicus)*, both male and female join in song, apparently sounding each other out for compatibility. These inhabitants of the dense forest canopies of central Africa look alike physically, as do all duetting birds. They open their strong, hooked beaks to voice clear tones that vibrate in their syrinx, an organ peculiar to birds which lies next to the trachea and allows them to articulate sounds without benefit of vocal chords. Although often separated by thick foliage, mated shrikes keep in contact by calling to each other antiphonally, the notes echoing back and forth between them so fast that observers cannot tell when one bird has stopped and the other picked up the tune, except from the printout of a sound-sensitive spectrogram. Female and male shrikes work out intricate combinations of notes that are original and unique to each mated pair and tell each the location of the other.

High grasses can be as dense as forests for small ground birds. The four-inch-tall warblers (Sylviidae) that skulk in the weedy plains of southern Europe also maintain contact by sounding long, integrated songs. Likewise, in Central America wrens (Troglodytidae) sing their melodies in a synchronized duet. Geography is no limit to this kind of communication. Birdsong can be tremendously complex, relaying more information than a female's or male's sexual availability. The birds apparently also inform each other where they are, and how much territory they control. Females know a good deal about their potential mates before they enter into what is often a lasting partnership. Moreover, some females choose their mates because of the appeal of the song. Laboratory experiments with lifetime-paired zebra finches from Australia show that a female seems to select her mate for his song, and after being separated from him for long periods of time will still respond to his song and select him again over another male.

Call is also the link between individuals in several primate species. Gibbons use songs to signal their location in the tangled southeast Asian forests. These small animals, along with siamangs, comprise a family (Hylobatidae) separate from the great apes but also removed from monkeys. They are reminiscent of *Homo sapiens* for their long, straight legs and overall body structure. Like some other primates, gibbons live in small family groups of an adult female and male and up to four youngsters. Unusually monogamous, the adults remain together for twenty or thirty years, their entire adult lives.

The gibbons' morning song usually lasts about fifteen minutes. The melody is complex, broken into male and female parts. Either sex can begin the performance, the first giving a series of introductory whoops and the partner continuing the phrase. Eventually the female utters her great call, a full voluminous series of notes that climb the scale to a great climax. The male tops this with a coda, shorter and sweeter but echoing the female's tone. From there the sequences repeat, each call lasting about twenty-five seconds with a moment of silence between broken by short whoops that mean the song is still in progress. Duetting gibbons send their melodies across a kilometer of forest. Other couples pick up the cue, stimulating one another until the forest rings from edge to edge for upwards of an hour as gibbons announce the dawn. They seem to select their mates for their abilities to sing together.

Sound travels underwater as well as through air, and the shallow-water croaker fish (Sciaenidae) use the strong muscles attached to the sides of their airbladders to send messages. Females apparently hear the Morse-code-like "boops" and swim to meet the males that mass together in their breeding season. Hydrophone operators in World War I discovered these small fish when they picked up their "boop boop-boop," suspecting the noise came from enemy craft. After the sailors identified the sound as the croaker's courtship, submarine commanders took advantage of them as natural camouflage. More recently, researchers at the National Marine Fisheries Service Laboratory at Port Arthur, Texas, have used hydrophones to pick up the courting song of the red drum, a variety of Sciaenidae that drums a tatoo before dusk. At this time the male's belly fades to silvery white and his back turns black, thus combining both sound and visual signals while the females remain bright red and silent.

Smell

Although scent, like sound, travels through water as well as air, most of the evidence in hand deals with atmospheric smells. Scent signals can be both remote and immediate, depending on the species that is sending them. Many female mammals emit special odors when they ovulate, and some species send out these pheromones throughout their cycle. But the intensity of the signals varies, so that some animals, like houscats, attract potential mates from what seems to be miles around, while others, like horses, check out the genitals by scent only as a last detail performed by a male who is already certain from visual cues that the female is sexually ready.

Scent signals are used by animals as diverse as moths and rhinoceroses. Rhinos are ungulates second only to the elephant in size among land mammals. Black rhinos *(Dicerus bicornis)* move about in small mother–infant groups on the southern African plains. When a cow is ready to mate, she forms a cautious but firm relationship with an individual bull. With a male following her, the female stops, then urinates, and continues to urinate at discrete intervals; there is an odorous substance in her urine that

A black rhino male (left) exhibits flehman beside a female and her calf.

informs him that she will be waiting at the end of the trail. So nose to the ground, the expectant bull moves from one scent post to the next, sometimes losing the carefully deposited trail and having to backtrack with his heavy body and try again. While moving from scent post to scent post, the bull twitches his lips into the "sneering" expression of *flehman.*

Occasionally an immature or almost mature female rhino approaches a bull and shows *prudity,* a term that incongruously describes an aggressive female's sexual curiosity. The action often irritates the males, but it gets those in the vicinity used to her particular presence. In some rhino groups, females and males seem to know one another individually and seek out company from time to time. They sometimes engage in what looks like a game of "chicken," rushing each other with horns raised, their nostrils snorting, only to stop short within inches of each other's face. The rhino courtship develops as the couple establishes a bond, remaining together throughout the day and sleeping curled up together at night, the bull's huge head pillowed against the cow's rump. If a predator threatens at this stage of their courtship, the bull will defend the female; at any other time it would be every rhino for itself. Rhino courtships may last a week or two, and are always initiated by the female's ritual urination.

Across the world in North America, smell helps the nocturnal cave-dwelling bats *Chiroptera* communicate in the dark. An enormous variety exist among these singular winged mammals, yet most share a propensity to live sexually segregated lives, large periods of which are passed in some sort of hibernation. A great deal is known about the biology of captive bats, but only recently have scientists begun to probe their lives in nature. The damp, dark places they tend to live in had frequently discouraged human researchers.

With part of the year reserved for sleep, courtship among bats is limited to a few weeks each autumn. At this time the female emits an odor from dermal glands on her face, chin, throat, or near her anus, depending on the species and where she is in her sexual cycle. When she releases her scent, the male bat approaches her in daytime, noisely purring and buzzing as well as emitting supersonic chitterings. This interaction goes on until they begin gesticulating at one other with their thumbs, forearms, and wings. Finally they seem, to the spectator, to embrace. All this activity takes place with the female perched upside down in her niche in the cavern ceiling.

Taste

Of all the senses, smelling and tasting most often work together. In primates we know that smell is used to modify taste. Observers in the laboratory have noted that tasting and touching frequently are paired signals in animals of widely differing species and for apparently differing reasons. Many cockroaches (Blattidae), those ubiquitous and often dust-covered insects, use no remote signals at all, but depend entirely on touch and taste. Adults of the drumbeater species *(Leucophaea maderae)* seem especially aimless and clumsy as they shuffle about awkwardly until a female and male touch, apparently by accident. Then, if the female is interested, she engages the male in an antennal "fencing match." Tilting her whole body toward his, she jerks her head with high-frequency movements. At the same time the drumbeater roach rocks against the hard ground, each drumbeat lasting about five seconds, the length of the pauses dependent on the female's receptivity. At the end of the fencing match, the male raises his wings slowly until they reach an arc of about 60 degrees. This signals the female to change her approach and court him now gently with her mouth. Passing over his exposed abdomen, she licks his body until she finds the glandular opening she is seeking under his wing. Here she feeds on his secretions, and in the sucking process stimulates him until he pulsates rhythmically, curving his abdomen upward. Occasionally the female feeds only briefly on this tergal gland, then moves on, leaving the male cockroach to mill about some more in apparent anticipation of bumping into a more willing mate. More often she continues until they copulate successfully.

Touch

Large mammals often follow a spate of urine-tasting with some form of body contact. Perhaps the male and female have been keeping company for days, gradually excluding other interested males as the courtship proceeds. Among elephants, when the cow is still in charge and setting the pace for the male to match, she prepares herself for consummation by using her trunk—a supersensitive extra "hand"—to reach under the male to caress his genitals, all the while rubbing her body all along his, apparently to keep him aroused.

Golden jackals *(Canis mesomelas),* canine neighbors of the Serengeti elephants, scavenge for food in large stable family groups. Paired monogamously, the jackals' "nuclear family" is part of a larger group that runs together—except while courting. Then the female and male part from their offspring and move alone, stopping frequently so that the female can groom the male's short fur with her tongue. For half an hour at a stretch she licks him, then settles back while he grooms her.

All male and female lions greet each other excitedly after each separation, even if it has been only for an hour or two, licking each other's faces and flanks and rubbing heads as if to reconfirm familiarity. While court-

Golden hamsters (Mexocricetus auratus) copulating, with the female in lordosis position.

ing, lions often increase the frequency of these gestures, an intensification of the everyday ritual.

Laboratory biologists have made extensive studies of another order's tactile approach — the rodent's. Female rats and hamsters have been studied for most of this century, and the female's unique behavior in initiating sexual activity has been analyzed and its evolutionary history puzzled over. The female initiates sexual activity by making sensuous body movements, raising her head and rump in a special posture. Eventually female rats reach a stage known as *lordosis.* At this point they arch their backs, a posture which facilitates copulation.

Just as lordosis is peculiar to these rodents, many female reptiles, such as lizards, respond to sexual overtures by adopting their own specialized posture to indicate sexual readiness. Studies of green anole lizards *(Anolis carolinensis)* from the southeastern United States show the importance of ritualized positions in their courtships. After an autumn and winter semi-hibernation under fallen logs and rocks, the male lizards emerge first and establish territorial control, frequently challenging one another with aggressive displays when there are few females about to join the audience. (Those females that surface early often witness these violent confrontations, and the exposure to such conflicts seems to hinder their own sexual development.) Later in the year when the male looks for a mate, he bobs his head. The female may initiate an exchange and bob her head; the male acknowledges her presence with a series of bowing movements. At the same time, he exhibits his great red dewlap, as proud an ornament, apparently, as the peacock's tail. The interested female lizard arches her neck toward him so that he can grab her by it, a gesture reminiscent of the neckbiting engaged in by members of the cat family.

David Crews' combined field observations and laboratory experiments with green anole lizards reveal a relationship between courtship behavior and the hormonal activity within the female just prior to copulation. The male makes the first move, bobbing his head toward a female. Later in the courtship the female may take the initiative, bobbing her head toward a lethargic male. But the female does not begin to ovulate without a male's active courtship gestures. His body movements stimulate her ovulation and have as much effect on her fertility as the outside influences of heat and light. She uses the same puppet-like head-bobbing gestures toward him as she would have him use toward her, until inevitably he picks up the cue and continues courting her on his own. The female's peculiar posture toward the male, and his toward her, contributes to building up the synchrony between them that is the height of courtship.

The degree to which females use signals actively to solicit male partners varies enormously from species to species, as do the kinds of messages they send to attract attention and indicate their sexual readiness. When female animals mature and eggs ripen inside their bodies, many females release chemical signals, fragrances that act as aphrodisiacs to the males of their species. These pheromones have been discovered in many but not all female animals. What seems to vary among species is the relative importance of these chemical signals to the other signs of courtship—sights, smells, sounds, and tastes. In some cases, as with the silkworm moth, a pheromone is the only signal needed. In other species it is but the first of a series of sensory advances which, taken together, form the rituals of courtship and lead ultimately to fertilization.

part ii

MATING

five

Reproductive Strategies

THE ADVANTAGES of sexual reproduction are scientifically controversial. Some evolutionists are puzzled about its existence, its origin, and its function. For whatever reason, logical or not, sex has evolved and apparently benefits those species that have it.

The main benefit of sexual reproduction came to be understood after the rediscovery in 1900 of the studies of the Austrian monk Gregor Mendel, who had demonstrated with pea plants that hereditary traits are transmitted by "factors." In the early twentieth century, Mendel's factors came to be understood in terms of genes, which carry the informational codes for offspring. The genes are located on chromosomes, threadlike structures in the nucleus of the cell. In sexual reproduction two mating individuals pool half their genetic material—half the chromosomes—to create an offspring. The new individual has it own unique genetic makeup (with a fresh combination of genetic possibilities that may help it to survive in the face of environmental change).

Let us assume along with Darwin that sexual variation permits a rapid accommodation to change of a cataclysmic type (though this is by no means a universally accepted explanation). For example, if the climate suddenly cools dramatically, those individuals with genes for thick fur will survive, while their less fortunate siblings will succumb. Over the eons glaciers have come and gone, the earth has shifted magnetic poles more than once, volcanoes have erupted, and islands have disappeared. It is probable that the animal life we know exists because sexual reproduction enabled new varieties to evolve.

Despite the evolutionary benefits, sexual reproduction seems like a process that goes against an animal's genetic self-interest. After all, animals that reproduce sexually pass on only half of their own genes to their heirs, while those that reproduce asexually pass on their entire genetic package. Evolutionary biologists who interpret success in relation to how much genetic material continues from one generation to the next are understandably perplexed by why sexual recombination itself ever evolved in the first place. Numerous single-celled species comprising an astonishingly large part of the animal world have maintained their genetic lines almost unchanged, enduring eons through periods of great change and continuing to thrive.

Sexual reproduction is a successful variation on, not a substitution for, asexual reproduction. It has been likened to the strategy of buying lottery tickets. The sexual animal buys lots of tickets with different numbers, while the asexual contestant buys one ticket and makes lots of photocopies. Variety in nature, though, does not guarantee survival.

Why Have Two Sexes?

Surely there must be some major advantage to sex. Biologists have found instances of sexual reproduction in life as simple as bacteria, which would make it appear that even these organisms find sex an advantageous way of propagating. But, on occasion, some animals as complex as vertebrates use strategies that avoid sexual reproduction. Females in a surprising number of species develop eggs into new organisms without the contribution of sperm. Whether these all-female species are a temporary adaptation or a permanent solution to survival in certain habitats is unknown. They do exist.

The development of eggs that have not been activated by sperm, a process called *parthenogenesis,* occurs most often among invertebrates and less frequently among fish and reptiles. The eighteenth-century Swiss entomologist Charles Bonnet discovered parthenogens while studying aphids. Parthenogenesis saves these creatures, which are under time pressure to reproduce, from having to find a mate. It does not help them, however, when their habitat alters, which may explain why aphids are heterogamous—that is, they alter their reproductive mode. They are parthenogenetic early in the season but reproduce sexually later on. True parthenogens reproduce only females of their species.

Darwin had information about gall-making wasps and saw-flies, both parthenogens, but dismissed the reports as dubious and gleaned from insufficient evidence. He suggested that the males of these species might yet be found. However, by the turn of the century other parthenogenetic species were identified, and it was no longer possible to regard them as an occasional aberration. In 1889 Sir Patrick Geddes, the British evolutionist, dismissed the phenomenon as "degenerative" but not mysterious. (When mentioning the occasional development of unfertilized frog ova in laboratories, he took the opportunity to add in a footnote that this was scientific evidence that a "virgin can conceive and bear a son" without its being a supernatural event.)

Species consisting only of females had been assumed to be limited to insects until 1932, when an all-female variety of fish was found in the Gulf of Mexico. Later, in 1958, the Soviet biologist I. S. Darvesky published his account of the first of what turned out to be at least nineteen species of parthenogenetic lizards. Other parthenogenetic lizards have been found in the southwestern United States and in Australia, which indicates that parthenogens are a widespread phenomenon. Darvesky had picked up only female specimens in the rocks and sand; so to test his suspicion that they were parthenogens, he collected the lizards *(Lacerta saxicola)*. He caged and lowered them into a crevasse over the winter to simulate hibernation. That spring the females produced eggs that hatched all-female broods. The next year, still in the lab, these lizards went on to produce yet another all-female generation.

Evolutionists have interpreted parthenogenesis as an optimal survival strategy in an inhospitable environment. When a lizard finds a new territory removed from other members of her species, it is more efficient for her to stake out a homestead than to go off in search of a male. Parthenogenesis saves the female from having to find the right mate at the right moment. At the same time, by producing females only, she does not risk creating competition from a male who would consume scarce food and produce no offspring.

Other females reproduce without fertilized eggs but rely on a heterosexual species to act as a catalyst. Amazon mollies *(Poecilia formosa)* are small fish that live in the waters of the eastern part of the Gulf of Mexico (and not the Amazon River, as the name implies). They need sperm to trigger reproduction, but do not incorporate any of the male's genes into the offspring. Some groups of mollies are all female. They depend on pirating males from neighboring schools of a different species. As the sperm approaches an egg, it releases an enzyme that causes immediate physical changes in the egg's membrane, which in turn trigger cell division inside the egg. The sperm itself never enters the egg, and all the offspring are female.

The molly's particular use of sperm—gynogenesis—is especially disadvantageous to the male of the kindred species. Although these males may prefer their own, they usually do not discriminate. In this event, all of the male's efforts leave him at the starting line in the race to procreate. Several close relatives of the molly actually use the male's sperm but

An Amazon molly, member of an all-female species.

produce zygotes—the result of the union of male and female sex cells—that are all female, discarding the male's chromosomes altogether. Mollies are a curious phenomenon that have been explained as a species in transit—in evolutionary time, a temporary species en route to becoming a new heterosexual fish.

That a sperm can initiate egg development, even when the spermal chromosomes make no contribution, is also illustrated in a wild strain of fruitfly (Drosophila melanogaster ms(3) K 81). This mutant strain includes defective, mostly sterile, males. When sperm from these males were used to inseminate females, the females produced viable zygotes, all but one of which contained no male chromosomes and were female. In this case, the males thus triggered what seems like parthenogenesis.

Why parthenogenesis and not androgenesis—the production of another individual from an unfertilized sperm cell? To determine whether the sperm is as mighty as the egg, researchers have removed the nucleus from egg cells and replaced them with sperm nuclei. The cells then developed into normal individuals, demonstrating what most biologists really knew already: that an egg is a wonderful cell indeed, because its very "egginess" includes the nutritional wherewithal to support the growth of a new organism. A sperm, by contrast, is mostly genetic material and perhaps small amounts of enzymes.

The Mechanics of Sex

Without exception, sperm by themselves are useless. And except for parthenogens, eggs alone will not produce a new generation. Thus it is necessary for both sexes that sperm meet and penetrate eggs, starting the process that results in offspring. The objective seems simple enough; its execution varies widely throughout the animal kingdom.

While females have sought to ensure that their eggs survive, males have sought to control the success of their sperm. Often the two goals are mutually exclusive. Members of sister species may employ different methods of fertilization, and certain reproductive organs have evolved in some species, only to disappear in others that seem to be more advanced in evolutionary terms. Even though a trend toward internal fertilization is undeniable, the evolutionary road is a zigzag one replete with cul-de-sacs.

Species sometimes arrive at seemingly dead ends that may prove entirely satisfactory for them.

How an egg becomes fertilized depends mostly on the kind of egg it is, and how many eggs the female has to release. Should a female produce a single egg every hundred years, she would take considerable care that it develop in the utmost security. Should she produce 10,000 eggs a month, she might simply release them into a promising environment where, according to the experience of her species (and the law of averages), at least one of her eggs would collide with a sperm and develop to maturity.

Many water-dwelling animals, or those that return to the water to reproduce, never mate at all in direct female and male copulation. They rely on *external fertilization,* where eggs find sperm outside the female's body. For species such as oysters, this is simply a fortuitous meeting of floating eggs and sperm. Most sea-living species do not trust to the random movements of the waves, however. Fish such as herring form a school; that is, great numbers of them swim together in a leaderless group. As soon as one herring spawns, the rest of the school, both females and males, release their gametes—eggs and sperm. It appears that the chemicals released by the first eggs stimulate synchronous emission by the rest of the school. With so many eggs and sperm released together, fertilization does occur, though with no selection of individual pairings.

External fertilization is common among fish and amphibians, whose eggs need moisture to develop. Only after an egg had evolved a shell to hold the developing embryo in its own liquid environment could egg-laying animals leave the crowded waters for the dry soil. But a few fish have evolved a unique alternative to a self-contained environment. The one-and-a-half-inch *Copeina arnoldi* of the Amazon deposits her eggs above the river on an overhanging piece of vegetation. The male helps in an unusual way. Both fecund female and mate leap from the water onto a leaf along the shore. There they lock fins, falling back into the water until the female has exuded a mass of half a dozen or more eggs. The locked pair clasp each other with their pelvic fins time and again and until all her eggs have joined together, and then she leaps back into the river. The male remains to cover the eggs with his sperm and then he stays on guard beneath them, splashing water up every ten minutes or so for at least three days until the moistened eggs hatch and the fry tumble downward into the stream.

Many sea-breeding females seek a male escort to ensure that their eggs get fertilized. An example is the Chinook salmon, which grows as long as five feet and may weigh a hundred pounds. Born in fresh streams in the Pacific Northwest from Washington to Alaska, salmon of this species spend the seven or so years of their adult lives in the Pacific Ocean. When the female is ready to mate, her ovaries grow heavy with eggs and she swims back, often against the current, to the fresh waters she knew as a fry. Male salmon fight the fast-moving waters, too, following her on the way to her homestream. When she arrives there, she sinks to the bottom of the river and digs a nest in the gravel into which she unloads her eggs. Immediately, before the waters can cover them with silt, the male enters her nest and fertilizes the eggs. Both female and male salmon make this exhausting journey only once. Thin and depleted when the spawning is done, they swim about in the river for another few weeks until they die. Ichthyologists have rescued some of these new parents and tried to revive them in captivity, offering them high-quality food to restore their strength. After spawning, neither female nor male has enough healthy cells left to respond.

For all its success, external fertilization is risky because fertilized eggs have high mortality. Some females must release hundreds or even thousands of eggs to ensure that even one of them survives. Fish that wrap themselves together into an "S" embrace lower the odds against all of their eggs drifting away untouched. An example are the tiny gray female pupfish that cavort with the dark blue iridescent males of their species in heated pools in the Nevada desert. Swimming side by side, they drop down to the algae at the bottom of the pond. Here they vibrate together, their bodies wrapped in an "S" until simultaneously they release their eggs and sperm.

Frogs have arrived at a different solution, called *amplexus.* The male, usually smaller, leaps atop a female and clasps her in a prolonged embrace — often lasting several weeks and occasionally going on for a month. The long embrace culminates when the female finally ejects a large glob of eggs, and he a cloud of sperm. This method guarantees more success than underwater spawning, though even amplexus releases unfertilized eggs to the vagaries of an uncertain environment because they may be preyed upon before the sperm find and penetrate them.

One adaptive solution to this risky interlude between release of eggs and fertilization is *internal fertilization.* The physical apparatus of the female's passageway and the male's penis had to evolve as well as new courtship rituals to assuage the animals' fears of body contact. Even after contact has been accomplished, and once sperm have found their way inside the female's reproductive tract, the next step is by no means either automatic or immediate. Some females store sperm, sometimes accumulating and keeping separate the contributions of several males before they allow the sperm to proceed toward the eggs. Females like the checkered

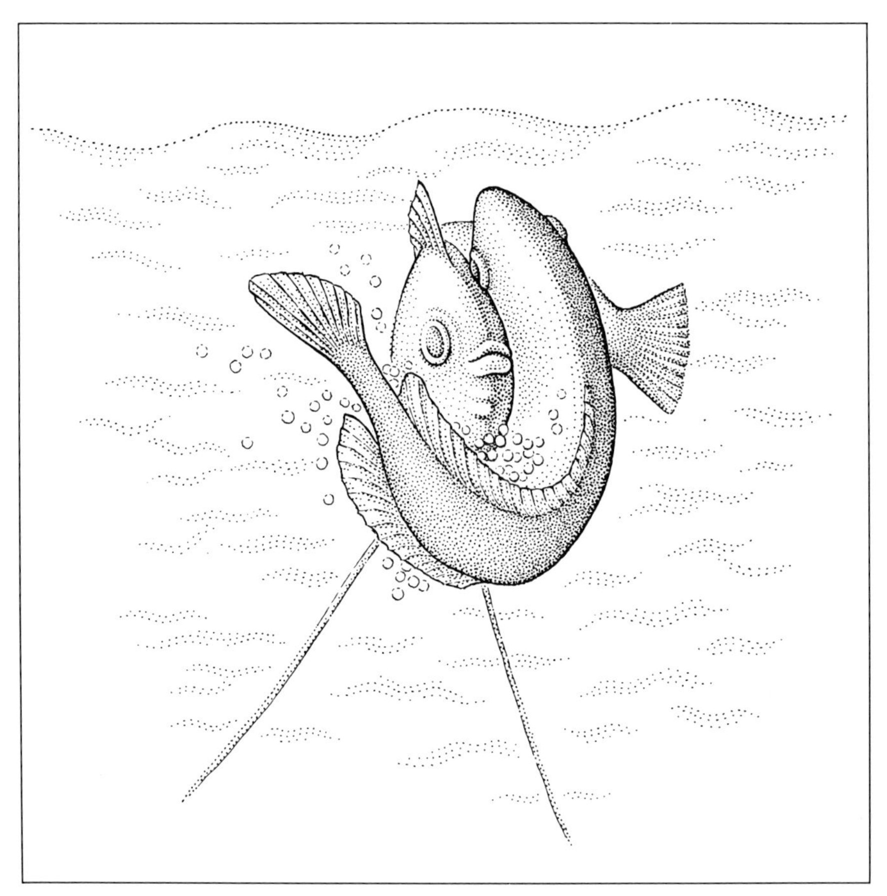

Some fish that fertilize externally raise the odds of success by adopting an "S" embrace.

white butterfly of Arizona, the southwestern mosquitofish, and the common fruitfly *(Drosophila)* accept sperm and then store them in special sacs. The butterfly keeps sperm for a few days, the fish for as long as ten months. Females of these species can thus produce several successive broods from a single mating.

These females may have to be inseminated only once in their entire lives, after which they can dole out sperm as they move about, fertilizing and then depositing eggs in safe niches such as on the leaves of wild mustard plants or under rocks at the bottom of ponds. When some butterflies run out of sperm, they solicit copulation again and refill their sacs, keeping within themselves the power to lay fertilized eggs at the right time and place.

Other females, including certain dragonflies, fish, frogs, turtles, and birds, allow the sperm immediate access to the eggs. This access is crucial to most reptiles and birds whose sperm must penetrate the egg before the shell forms. What happens to the eggs next is a question of nurture, for the mating operation is by definition complete. These egg-laying females are called *oviparous* because their fertilized eggs develop outside of their bodies.

In a harsh environment, other egg-laying females, including both those that copulate and those whose eggs are fertilized externally, reclaim the fertilized eggs and guard them inside or atop their own bodies. The African cichlid *(Haplochromis burtoni)* lays eggs and then carries them in her mouth. The male of her species sports decorative spots on his rear fins that seem to mimic eggs. When she spots him, she opens her mouth, as though trying to pick up any eggs she might have overlooked. The male takes advantage of the occasion to squirt his sperm inside, thus fertilizing the eggs. Females of another freshwater fish family, the Amblyopid, place fertilized eggs inside their gill pouches until the young emerge weeks later, miniatures of their parents. Still other females, such as the gastrotheca frog, transfer their fertilized eggs to maternal pouches on their backs (despite their name, which indicates a special stomach). The pouches look like sores but are really nests for her eggs.

Common North American rattlesnakes, in the Crotalidae family, retain fertilized eggs within their bodies and give birth to living young, as do boa constrictors (Boidae). And a few female animals, such as the cartilaginous nurse shark, give birth simultaneously to both live young and eggs that will not hatch for several days. This species seems to underscore the point that there is no "best" way to reproduce, even in a particular animal.

The nurse shark that gives birth to living young, as well as the rattler and the boa, are *ovoviviparous;* that is, they retain the fertilized eggs within their bodies as a better, super shell. Unlike truly *viviparous* animals, such as ourselves, whose young are nourished by their mother as they grow inside her, these eggs develop independently of their maternal environment. They could as easily be carried about by their fathers—and indeed that does occur in the male midwife toad, a small European native which carries strings of fertilized eggs wound about the lower half of his body, and the male pipefish, which nourishes the embryos.

Eggs appear in their full complement at birth in all female mammals,

The male African cichlid sports "egg" spots on his anal fin, which attract the mouth-brooding female.

birds, and many reptiles. The animals are born with tiny egg follicles not ready for fertilization until the females mature sexually. At maturity, depending on the species, a single egg may be released, or a whole mass may be ready for fertilization at once. In insects and fish, on the other hand, the female continues manufacturing eggs most of her life. The queen bee imbibes royal jelly and produces vast quantities of eggs for several years. Female fish like the carp continue growing all of their lives; as adults age, they produce many more eggs than when they first matured.

All eggs develop in ovaries, and they usually need some passageway, or duct, through which they can leave the ovary when they are ready and move into another chamber within the female's body, or into the outside world. In birds, reptiles, and monotremes—an egg-laying branch of mammals which includes the platypus—the eggs share this passageway (known as a cloaca) with excretory substances.

In reptiles, birds, and mammals a gland in the ovary called the corpus luteum is activated after the follicle has ruptured and released a ripe egg. The corpus luteum secretes a substance that forms a hard shell on bird and reptile eggs, but in mammals it secretes the sex hormones estrogen and progesterone, which stimulate the thickening of the uterine lining so it can receive the embryo.

Most female mammals also have a clitoris, an organ that is a homologue of the male's penis. Its function in some species is unclear, but in primates there is little doubt that the clitoris provides sensations that incline the female to seek out sexual intercourse. Mammal females have vaginas that are simple tracts or divided passages through which the male injects his sperm; in a divided vagina he uses a divided organ of his own. This vaginal passage leads, in mammals, to a flexible organ, the uterus, which comes in a variety of shapes and is sometimes divided into two parts, or horns. The uterus is the protective case where young can grow.

Some female mammals mature with a membrane over their vaginal canals that must be pierced or stretched during sexual intercourse so that the sperm can travel to the eggs. In female humans, the membrane, called the hymen, is broken during the first sexual intercourse, if not earlier. In the great fin whale it is not completely destroyed until childbirth rips away the thick membrane. Moles, mole-rats, and hyenas all have hymens that must be ruptured during their first copulation. Afterward, the nonvirgin female is recognizably changed. The virgin female mole has a pseudo-penis, an elongated clitoris containing a closed vaginal opening like the hyena's. Thus the virgin female mole's genitalia look very much like the male's (and since moles do not live very long, English farmers used to think that female moles appeared only in spring). Tiny nocturnal prosimians (Lepilemur) in Madagascar also sport vaginas that open only during the mating season. The rest of the time the vagina is completely sealed with tissue, making sexual activity impossible.

Such adaptations are explicable only if the male of the species finds it to his advantage to seek a virgin. But there is no evidence that mammal males seek inexperienced females, and no evidence that females with this peculiar anatomical feature remain monogamous. Among mole-rats, a bald, subterreanean colony-dwelling species in Kenya, Somalia, and Ethiopia, the fertile females develop perforated vaginas that close up if the females are not impregnated. Yet in whales, one can explain the resealing of the vagina as a means of keeping water out of the reproductive organs.

Sex among the Hermaphrodites

All females did not evolve along a single path. Many are born with the gonads of both sexes and live sequentially as one sex at a time. Like Tiresias, the legendary Greek who lived ages as a woman and then as many years as a man, these individuals—*hermaphrodites*—alter sex according to an unknown mechanism that may be triggered by light, or temperature, or even social pressures that stimulate hormonal changes.

In the shallow waters of the Australian Barrier Reef, small schools of cleaner fish *(Labroides dimidiatus)* slip in and out of the coral catacombs, foraging for food. Ross Robertson, an American ichthyologist, has observed the fish as they scout in small groups, all but one member of which are female. At first glance their lives seem hazardous, for they feed on parasites that live inside the gills and mouths of large sea bass. Aside from size, cleaner fish look alike. All the females lay eggs fertilized by the unique male who leads them through the waters, attacking them with aggressive displays as if to prove his strength. From time to time he disappears. And then, within the hour, the largest female begins to flaunt her strength, putting on the same kind of aggressive show that the departed male had recently displayed toward her. As her behavior changes, so does her sex. Within hours she is leading her group. Two or three weeks later "she" releases sperm and is, in effect, a male. But should the vanished male return, the newest male calms down, stops producing sperm, and produces eggs once more. He is a she again. Observers have noted individual fish oscillating behaviorally between sexes three times before settling into a permanent role.

Another kind of hermaphrodite lives in colonies inside the shells of hermit crabs off Cape Cod. A tiny mollusk, the slipper limpet seems asexual in infancy and will grow into a female or male, depending on which adult of its species it chances to settle next to. The presence of a

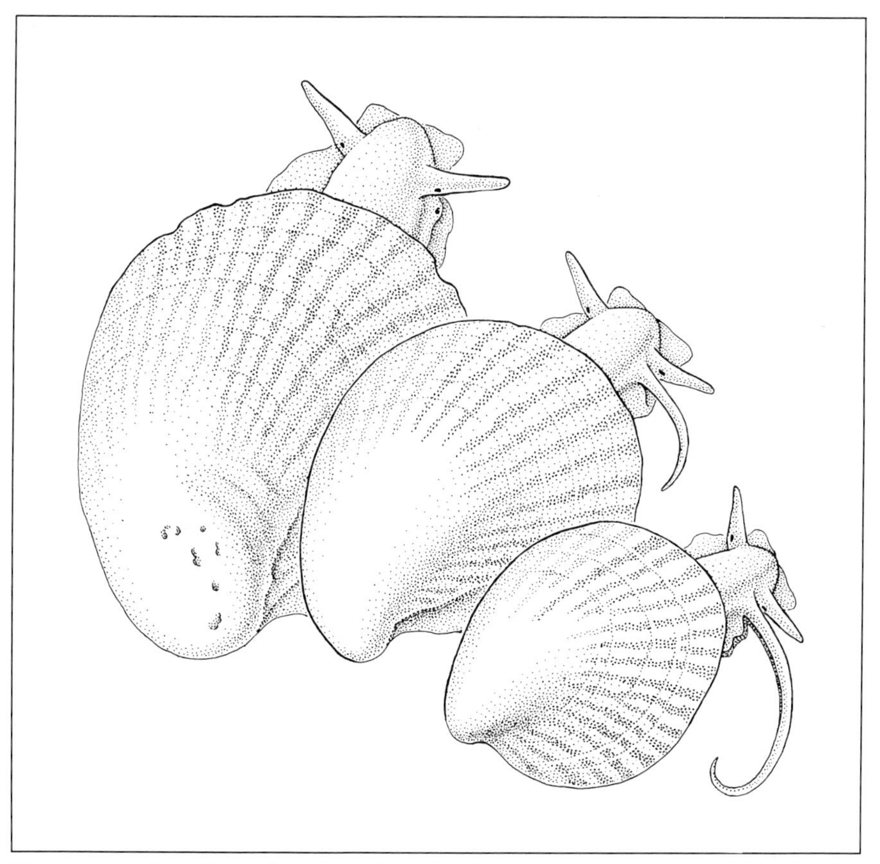

The progression of sex change in the slipper limpet, from male (right) to female.

female makes it become a male, and vice versa. Its sexual identity need not be permanent. Eventually every young male limpet loses his sex organ and becomes female. Limpets that grow up in isolation usually become female, however, and switch to male even at a mature stage if an adult female appears on the scene.

Still underwater, *Bonellia* offers a Tiresias-like experience to only half its offspring. All of these sand-dwelling worms begin life as free-swimming female larvae. Those that remain female mature and eventually burrow into limestone crannies where they start reproducing. A variety in the Bay of Naples inhabit tunnels in huge chunks of volcanic rock. In such dwellings they continue to grow to several inches, not counting the additional length of a long proboscis nose which can extend as far as six feet into the waters. With its nose the adult female fans the water and picks up larvae. Those that avoid her whip remain female. The larvae that do not, adhere to her nose and are there transformed to adult males, tiny creatures only a fraction of their host's dimensions. They soon atrophy, remaining first on the female's nose and then gradually moving into her mouth and through

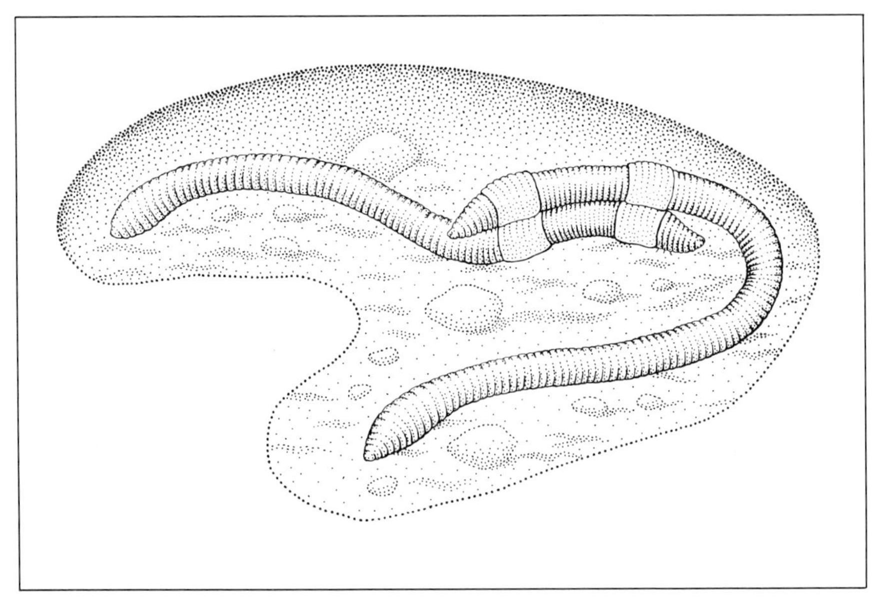

Hermaphrodite earthworms engage in copulation.

her body to her oviducts where they fertilize her eggs, setting loose a new generation of female larvae that will swim the same gauntlet themselves.

This kind of sexuality, termed labile sex determination, is a form of sequential hermaphroditism. That is, individuals never exhibit both male and female characteristics at once, and the trigger for the hormonal switch is the presence of an individual of the opposite sex. Among species that take turns at being female and male, sex does not appear to be determined at conception by a single chromosome, which is the way it works in warm-blooded vertebrates. In mammals the presence of a Y chromosome determines that the embryo will be male, and in birds maleness is made certain by the presence of an X chromosome. Yet in various reptiles the complement of chromosomes does not seem to determine sex one way or the other.

Reptile studies reveal that the sex of turtles is determined by the temperature at which the eggs incubate. A difference of 6°C affects the zygote in that higher temperatures produce females and lower ones males. Under normal conditions the sexes remain balanced by the judicious laying of eggs. This sexual ratio could become unbalanced should the climate suddenly alter for a season or two, thus wiping out one sex or the other from the population. This phenomenon has been observed in several species and is known as environmental sex determination.

Some hermaphrodites carry both genitalia simultaneously. The common earthworm, for example, always needs a double partner. The two individuals line up together so that a special organ, the clitellum, secretes mucus that holds them together as they pass sperm into each other's receptacles. After a few days the mucous package slides forward and the sperm meets the eggs. The package continues moving and carries with it the newly fertilized eggs as it slides completely off the body of the worm. On the ground the mucous makes a kind of cocoon atop the earth, from which minute earthworms emerge.

Transferring Sperm to Egg

Male earthworms simply manufacture sperm and accumulate it in a chamber which they press against the parallel female opening. The system is not very much more complex in most birds, where the sperm accumulates in the male's cloaca, which he in turn presses against the female's cloaca. Thus he depends on the sperm's own power or the female's vaginal muscles to carry the sperm along toward the eggs.

But males of many species insert their sperm directly inside the female by an assortment of appendages that zoologists describe generally as *intromittent organs,* saving the term *penis* for the sperm-conducting, erectile organ of mammals and some reptiles and birds. What distinguishes a "true penis" from the analogous organs of insects and fish seems to be its evolutionary and embryonic antecedents, not its function, which is simply that of transferring sperm internally to the females of its species.

Where to insert the sperm, and with what, have been solved by several curious adaptations. All male animals manufacture sperm as soon as they mature sexually, and they continue its manufacture throughout their lives. Some produce spermatozoa in a liquid environment, expelling the surrounding fluid into the female along with the sperm, and sometimes providing an additional source of nourishment for the developing embryo. Others emit sperm inside a hard case, a *spermatophore,* which may serve as a plug to prevent leakage after insemination, as with spiders, butterflies, rats, and chimpanzees.

A few grasshoppers and butterflies, like worms and most birds, place their sperm directly inside the female genital opening without a special intromittent organ. The males press their genital openings against the female's genital opening; the sperm move into and through the female's reproductive tract by muscular contractions. Wasps and bees, by contrast, have organs analogous to penises, which force the sperm directly into the female's vaginal opening. She, in turn, receives and stores them in her bursa copulatrix, a special sperm-storage area, for use at the appropriate time.

Wolf-spiders *(Lycosa pseudoannulata),* along with most other arachnids, never develop a penis. Instead, the male of the species uses several of his eight appendages as secondary sex organs. Three pairs of legs and a pair of pedipalps, equipped with tiny hooks or spines, play a double role: they enable the male to hang onto the female while he orients himself to deposit his spermatophore inside of her; and, with sperm in hand, so to speak, he uses his pedipalps to insert his sperm into the pair of openings in the female's epigynum, a special chamber in front of her vagina that stores sperm. Some male spiders fill one opening at a time, while others manage to use both pedipalps to fill the female's two openings simultaneously.

In the insect world, a variety of alternatives evolved that seem as effective to their users as a mammalian penis. The common bedbug, of the Cimicidae family, is a small, flat, wingless creature that inhabits most parts of the world, especially where humans reside. The male wears a spike like a hypodermic needle in front of his penis. When the female bedbug happens to bump into him, he stabs her in the back and his sperm enter her blood stream and travel through it to her eggs. The noted German zoologist Wolfgang Wickler suggests that ancestral bedbugs probably copulated using intromittent sex organs, and that the aberration arose by chance. It remains, he suggests, because it works, and so was not removed by natural selection. Idiosyncratic as it is, it persists because bedbugs proliferate well this way.

Still other invertebrates like the dog leech *(Erpobdella),* an annelid that lives in the guts of dogs, have their successful methods of transferring sperm to egg. The female dog leech seems to sense the sudden arrival of a foreign substance somewhere on her skin. It is her male's spermatophore,

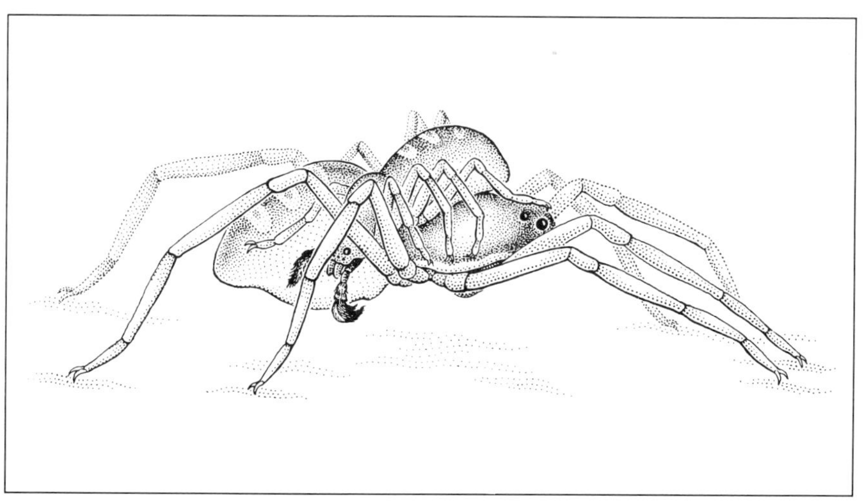

The female wolf-spider accepts the sperm-bearing pedipalp of the male in her epigynum.

a foreign body that triggers an infestation. Soon the blister where the sperm has landed develops into an open sore. The female's resistance cells combat the infection when the sore opens and thus the sperm seep in, entering her blood system where they swim about until they reach her ovaries. This is not a unique situation. The velvet worm, a land dweller of the *Peripatus* genus, is also infected by its male, which sets up a similar infestation.

Octopodes *(Octopus vulgaris)* are another invertebrate, not fish at all but rather large cephalopod mollusks. With eight sensitive tentacles, they never apparently needed to evolve a special organ to transfer sperm. Instead, one of the eight tentacles functions as a spermatophore inserter. In a manner somewhat akin to the spider, the female octopus allows the male of her species to insert his sperm, by "hand," into her appropriate orifice.

The female octopus accepts the male's tentacle, with which he inserts his spermatophore into her orifice.

Evolution of Sex in Vertebrates

The first vertebrates appeared in the water, and most that remain there still rely upon external fertilization because their eggs survive in an aqueous environment. Copulation and internal fertilization in fish are more than a curious anomaly, however. Indeed, male sharks and rays have intromittent penis-like organs. In addition, all male cartilaginous fish develop copulatory appendages called "claspers," with which the males hold onto the female's body. In some species these come in pairs and are used, like the male spider's pedipalps, to insert sperm into the female's genital tract. Some kinds of skate, such as the Texas skate *(Raja texana),* have erectile tissue in this organ. These cartilaginous species fertilize internally while most bony species of fish do not. These particular adaptations are evolutionary dead ends. Internal fertilization is not necessarily an "advance" in terms of the individual organism.

Some teleosts, as bony fish are termed, have an enlarged anal fin that is both rigid and movable. Called a gonopodium, a quadi-penis carries sperm directly inside the female's opening, where they fertilize her eggs. The cusk eel is one such fish whose eggs are fertilized and develop completely inside the female's body; the young are born alive.

In other fish the long anal fin is modified into what is called a priapium. Tiny Malaysian varieties of the family Phallostethidae, a 15-centimeter denizen of the Malay Peninsula, have a priapium near the head and shoulder that contains the anal opening as well as the urogenital outlet. It functions as a clasper as well as a sperm duct, allowing the male to hold tight to the female, whose genital opening is located between her pectoral fins. Slightly different is the arrangement of the family Horaichidae that lives in the fresh waters south of Bombay, India. The anal fin has evolved to transmit sperm, but in a packet similar to the spermatophores of insects.

The male intromittent organ is an optional apparatus in fish, as contrasted with the female's reliable opening. Although some males do have some kind of sperm inserter, the appendages appear less frequently than they do in most terrestrial species.

As animals left the sea, amphibians continued their predecessors' practice of external fertilization, with the exception of some salamanders and one species of frogs. The *Ascaphus truei,* or tailed frog, of the Pacific northwestern United States lives in swift icy mountain streams. This frog has lots of blood vessels beneath the skin covering both the female's and the male's cloacal openings. In the male, the blood vessels cause an erection, which turns into an intromittent organ; so the female gets her eggs fertilized internally.

Reptile species like turtles and crocodiles sometimes include a small clitoris in the female, and always a penis in the male. All oviparous reptiles lay amniotic eggs, with thick leathery shells, which must be fertilized inside the female's oviducts. In snakes and lizards the penis is split, or double, and are known as hemipenes; only one penis at a time is inserted into the female's cloaca.

A female platypus curls around her eggs, awaiting the birth of young that she will suckle like an ordinary mammal.

If evolution followed a straight line, it would seem likely that the next great change, warm-blooded birds, would sport these organs too. As with the boneless fish, though, only the most primitive birds such as water foul, ducks, and geese, the ostrich, and many flightless birds, have penises. As birds began to stay aloft in flight for longer periods of time, species evolved that seem to be streamlined for speed. Females make do with a single ovary, and male birds have lost the penis altogether. Relying on an efficient meeting of cloacas, birds enjoy the benefits of internal fertilization.

But evolution seems to have followed branching lines. Monotremes, the small subclass of mammals containing Australia's spiny anteaters *(Tachyglossus aculeatus)* and platypuses *(Orinthorynchus anatinus),* act in most ways like viviparous mammals, suckling their young after hatching them from eggs. But sexually they perform like reptiles. The platypus's cloaca is divided so the sperm have a separate channel, quite apart from the excretory passages, and the penis itself is forked, to match the female's forked vagina. Marsupials, the mammalian subclass that includes kangaroos, koala bears, and opossoms, also have this odd forking.

Although female and male sex organs do not follow a single line of development, internal insemination and the equipment to arrange it appears more commonly among the more recently evolved forms. Internal insemination prevents the wastage that occurs when eggs and sperm meet outside the female's body, and thus individuals of both sexes can produce fewer gametes. Consequently, both parents are more careful about each copulation. Fewer offspring will be born because fewer eggs are produced, and it is to the benefit of both parents to see that each sexual encounter is fruitful.

The independent adaptations of females and males of a single species cannot be so different as to prevent mating. There has to be a certain anatomical compromise. The earthworms that crawl about gardens can only copulate with other worms of the same length. A great part of their lives can pass as they line up with one worm after another, looking for the linearly perfect mate. Another example of anatomical compromise swims in the coastal waters of the Pacific Ocean in Central America and in some freshwater streams as well. The often foot-long viviparous four-eyed fish *Anableps anableps* lives on the surface of the sea and has evolved two pairs of eyes, each with two corneas and retinas (for seeing in air and for underwater viewing). This female has a peculiar problem: The male of her species has a penis-like organ that is able to move either to the left or the right side of his body; she has a scale that covers up her genital opening on one side too, so it takes a "left-penised" male to copulate with a "right-vaginaed" female, and vice versa. The left–right ratios are not equal in the populations of these fish, which might account for the low population if we assume that some individuals never find a mate.

Under Lock and Key

Female animals produce eggs of varying complexities, and some species have partial control of egg production. Males, in evolutionary response, evolved new methods of insemination, in part to remove some of that

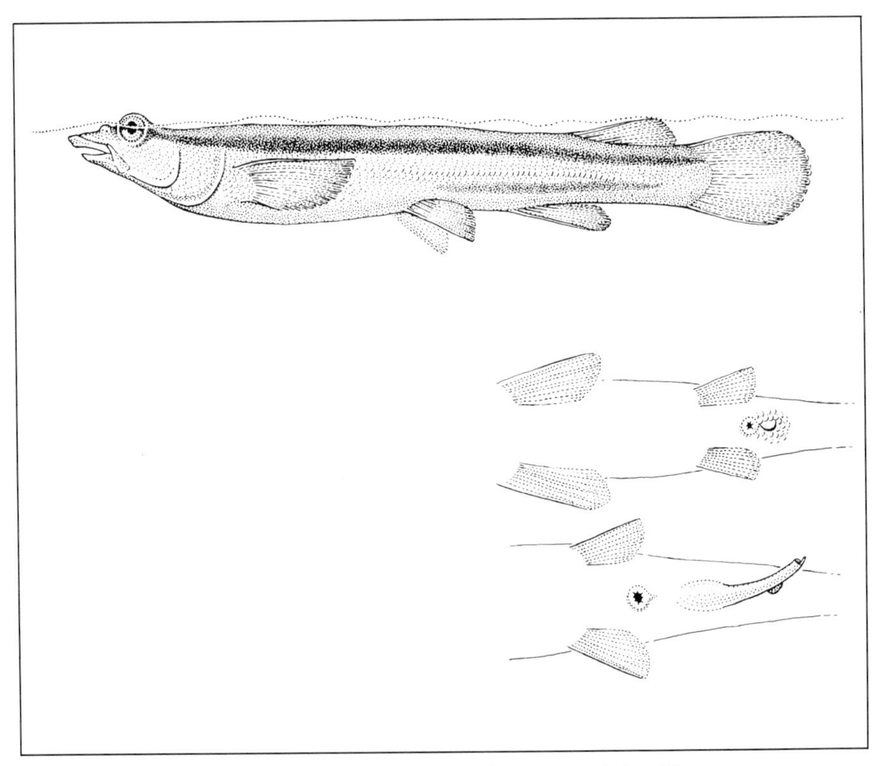

The specialized left- and right-sided sexual organs of Anableps.

control from the egg-bearers and retain it for themselves. Thus mating is the opportunity for each partner to exert what influence it can on the offspring. Both appear to want the last word on determining who the father is.

Some biologists use the metaphor of "lock and key" to describe the way genital organs fit in species like the four-eyed fish or spiders. This perfect match, which automatically prevents interspecific mating, satisfies a logical sense of order. Not only does the lock ensure that fertilization takes place, but it prevents the sperm from leaking. From the male point of view, copulation includes the injection of sperm into the female as well as an effort to see that the sperm remain inside long enough to meet the egg.

The lock and key that hold spiders together may be more comfortable than the adhesive excreted by certain male parasitic worms. These males have cement glands alongside their testes which prevent the females from moving away until they have completed intromission. Other parasitic worms stay together for their lifetimes, which among the schistosome is as long as thirty years. In these three-quarter-inch parasites the slender female fits into a deep groove in her mate's underside. So joined, she produces around 3,500 eggs each day in a monogamous relationship that is without equal.

Copulation appears to play a psychological role between animal pairs that remain together. It is the occasion for couples to become intimate, even though they may not achieve fertilization at the first try. In some monogamous species, like mallard ducks, it is likely that copulation becomes part of the bond. Mallards frequently copulate before the female's ovaries produce eggs. It is as if such intimacy ensures ease and familiarity so the act will be successful at the right time.

With the appearance of sexual matters as the subject of open—if not obsessive—discussion in recent years, a new mythology has blossomed about the position in which *Homo sapiens* performs sexual intercourse. Among these myths is the assertion that humans alone copulate face to face and for an extended period of time, achieving an intimacy unknown in the rest of the animal kingdom. Yet primates other than ourselves indulge in lengthy sexual intercourse during which they face each other. Especially interesting to evolutionary biologists is the behavior of the rare pygmy chimp. This small ape, a separate species from the larger chimp, has been singled out as the nearest animal alive to what many evolutionists see as our hominid ancestor. Pygmy chimps perform well on the kinds of intelligence tests primatologists have been administering to apes for half a century. Their sexual intercourse is marked by a great deal of eye contact and mutual patting in this face-face-position. Equally interesting is the fact that pygmy chimps indulge in intercourse when the female is not ovulating and cannot become pregnant. They copulate very often, as a greeting gesture, giving the impression that they are using copulation as a tension-releasing mechanism.

Orangutans also face each other during sexual intercourse but, unlike chimps, do so while hanging from the upper branches of trees in the forest canopy. Biruté Galdikas, a primatologist who has been studying orangutans on the island of Borneo since 1973, reports that she has seen intercourse preceded by lengthy oral contact by the male with the female's genitalia. The act itself can last as long as twenty-eight minutes, during which the male grumbles and the female remains silent.

Most often, however, throughout the animal kingdom, from insects to baboons, the most common position for sexual intercourse has the female standing or crouching, presenting her vaginal orifice to a male who approaches her from behind. Exceptions abound here, too. Among birds, penguins in the Antarctic occasionally reverse positions, the female rising to place her cloaca atop her mate's. Pigeons also alternate position, which may simply indicate that the sperm in these species is vital enough to swim upstream, or that the female's urogenital tract has muscles strong enough to contract the sperm upward toward the eggs in her ovary.

Unique is the mating posture of female cats. Instead of standing upright while copulating, they lie with forequarters flat on the ground. This forces the male to mount with his forepaws resting on the female's shoulder, supporting most of his weight. R. F. Ewer speculated that the prone copulatory posture is the one in which the male, balanced as he must be to reach the female, has the least leverage. Reminding us that cats are among the world's most efficient killers who often destroy their prey with a single bite to the neck, she has speculated that this somewhat awkward position makes it difficult and perhaps impossible for the male to confuse the female with potential prey. Other zoologists suggest that cats are well aware of who their prey is and use this position simply because it is the easiest way to support the male's weight.

Although ground level is the most common site of copulation, a few avian species, such as swifts, meet in the sky and copulate on the wing. Copulating aloft has also been perfected among honeybees, whose queen accepts the spermatophore of many drones while buzzing several meters above the ground. Aloft, but not in flight, the American opossum, a small furry marsupial that flourishes all over the continent, copulates upside down, hanging from a branch. And the sloth, an inhabitant of the rain forests of Central and South America, also copulates upside down, too, as do most bats. Horseshoe bats, for example, mate as if standing on their heads, inside the dark caverns of Carlsbad, New Mexico, where thousands of them hibernate in winter.

Those fish that copulate necessarily do so under water. But so do several mammalian species, including the hippopotamus *(Hippopotamuses amphibius).* The enormous hooved African mammal seeks partial submergence. The largest animals in the world, whales, court each other with displays of leaps and splashes. Then they settle into a side-by-side position in which the male erects his ten-foot-long penis from the slitted pouch that normally secures it, for streamlining, in front of his navel. The few observers who have seen whales mate report them copulating while leaping in the air, or lying belly to belly in the sea.

Sexual intercourse can last half an hour, as leisurely as with an orangutan, or barely seconds, so brief an exchange that with hummingbirds an observer can be sure of the act only by capturing it on film and reviewing the episode in slow motion. Length of intromission is independent of the species' size, or its taxonomy. The orangutan's closest relatives, chimpanzees and gorillas, indulge for just a few minutes, as do many antelope. Butterflies stay locked together for more than an hour. Members of the canine family, including wolves and dogs, are also bound together during copulation. In their reciprocal sex act, the base of the male's penis expands while the female's sphincter muscle contracts, locking the male in place for as long as an hour. Amplexus in frogs can last for six months. And for the schistosome, in which the male is fused into the body of the female, the connection is permanent.

Frequent intercourse sometimes substitutes for the large number of thrusts a female must experience during a single encounter before the male releases his sperm. With rodents like rats, squirrels, and beavers, the male must thrust several times before he can achieve an ejaculation. Felines have the same demands. With the male bee, his one effort is enough and proves to be both culmination and termination of his short life.

The location of a mammal's vagina has a lot to do with the nature of the copulatory act. In elephants it is in the middle of her belly, in front of her rear legs, necessitating strategic maneuvering from the male, whose lengthy penis is equipped with extra muscles that allow it to bend twice, like a plumber's elbow, to reach back, and underneath, then up into the vagina.

Some female spiders fall into what one entomologist describes as a cataleptic state throughout the duration of copulation. Other females, be they monkey or fly or bird, though fully conscious, seem to be inattentive to the act of intercourse, munching a delicacy provided by the male. Still others appear to experience a sensual and physiological climax, which they express either vocally (as with macaques, baboons, and chimpanzees who grunt, shout, and send out a call at the apparent moment of climax) or with apparently involuntary pelvic movements, lip-smacking, and clutching backwards at the male with alternate hands—movements that the anthropologists Doris Zumpe and Richard Michael dubbed a "clutch reflex" when they observed it in laboratory rhesus macaques. Some primatologists speculate that the female's cry or clutchings actually trigger the male's ejaculation.

Apparent pleasure is not the only cause for noise-making. Female elephant seals and Barbary macaques emit copulatory screeches. With both species, females seek multiple male companions. It seems that they literally call attention to themselves and their apparent interest in the sexual act in order to attract new partners who will interfere, giving the females additional chances of becoming pregnant.

The "love bite" is another familiar expression of fulfillment. Crocodile females swim to a mate, allowing him to rub his throat against her snout, releasing a musk-gland secretion. Then, as he penetrates her body she allows him to take her throat into his jaws in a mock bite. In like fashion the female chuckwalla *(Sauromalus obesus)* lizard of the southwestern Arizona desert lies still so the male can take her head into his jaw. In many shark species the female allows the male to bite her back and pectoral fins; a mature female's back is often a welter of criss-cross scars. Head biting and neck grabbing is also common in mammals including almost all members of the cat family—tigers, cheetahs, and panthers—and elephant seals.

The moment of sexual excitation passes, and in some species the sexual act seems quite forgotten. Abrupt separation is not universal, however. A postcopulatory involvement occurs among wild African zebras *(Hippotigris),* who continue to groom each other after mating, extending the intimacy beyond the quick necessity of copulation. With Uganda kob, the lek-courting ungulates of central Africa, with whom copulation takes just seconds, the postcopulatory ceremony lasts as long as five minutes. The female stands still, her back legs spread, while her sexual partner licks her vulva, her head bent in a swanlike pose. H. K. Buechner, a German ethologist, suggests that this nuzzling provokes her glands to release oxytocin, a hormone essential to uterine contractions that transport the spermatozoa to her oviducts.

Remote from these mammals, the bubblenest-building fish Anabantidae in tropical streams perform a postspawning ritual. Both female and male remain immobile as the eggs sink. After a few minutes the female swims away while the male stays to retrieve the sunken eggs with his mouth, which he then spits one by one into the safety of his nest. The small damselflies *(Calopteryx virgo)* of the eastern United States dance a postcopulatory chase in which the male follows the female, guarding her so she can lay her eggs uninterrupted, preventing his sperm from being displaced by another ambitious male of his species.

Postcopulatory rituals demand two players, an impossibility for some species. The polychaete *Neanthes arenaceodentata,* a small worm, is devoured by her mate almost as soon as she has laid her eggs. The situation is reversed for some varieties of mantis *(Mantis religiosa)* and for black widow spiders *(Latrodectus mactans).* A female mantis may spend five or six hours coupling with, and usually attempting to eat, her mate. Beginning with the male's head and neck, she devours almost all of him while he is still sending sperm into her body; eventually, only his wings are left. She may mate again although she is already fertilized, and—if she can—she will eat this second mate as well. In the late nineteenth century, Henri Fabre observed a caged female mantis's consumption of seven males in two weeks. Thus, no matter how many males she mated with, she was always without a partner for any postcopulatory rite.

The mechanics of fertilization, both external and internal, have clearly evolved reciprocally—the key must always fit the lock. Whatever method the female has evolved for producing eggs has evoked a satisfactory response from the male of the species.

six

A Question of Timing

LIFE ON EARTH is cyclic. We are all responsive to the movement of the moon around the earth, or the earth around the sun. The moon's gravitational pull upon the oceans results in the tidal changes that affect marine life, especially those animals that live in the shallows of the reefs or the continental shelves. For females in most species, both marine and terrestrial, the new structures and hormonal systems that develop as they reach maturity function only when the time is ripe. And that time is determined in part by the length of daylight or the brightness of the moon. As females respond to the sun and moon, so must the male of the species respond to them.

In springtime, when the moon is full, horseshoe crabs, armored arthropods that have existed unchanged for at least 175 million years, crawl onto beaches on the Atlantic seaboard. Here they mate and bury their eggs under the moist sands. At almost the same time of year the grunion run in California. Just a few days after the full moon when the tides are high, the Pacific beaches host silvery-scaled fish. They leap from the sea en masse to lay their eggs at the inter-tidal zone, so the newly hatched grunion will be flooded one month later when the tide is high again. Like these grunion, an extraordinary proportion of all animals are nocturnal. As night creatures, they live in rhythm with the stages of the moon, its light and shadows. In the West Indies, female marine worms perform a "nuptial dance" keyed to the phases of the moon before they expel their eggs into the sea. Then the males of their species who had watched the performance respond by ejaculating to fertilize the freshly spawned eggs.

In ways that are not completely understood, the moon affects the reproductive lives of terrestrial diurnal animals, too. Ancient humans noted the correlation between the human female's menstrual cycles and the lunar cycle. In historical time, zoologists have observed that females of many species, including langurs, lizards, gorillas, and cockroaches, ovulate spontaneously about once a month. The human female usually releases one egg at a time each month, which moves through the oviducts to the mouth of the uterus. If the egg is not fertilized during the course of this journey, it disintegrates. And at the end of the cycle the uterus sloughs off tissue it had held for about twelve days in preparation for the arrival of a fertilized egg. Menstruation is, in fact, a sign of infertility. Old World monkeys and apes menstruate too, though none with as marked a flow of blood as the human female.

Unlike humans, most mammals follow an estrous cycle that draws attention to the hours or days surrounding ovulation. In many species this is the only time in which the adult female is sexually curious and receptive. An estrous cycle is divided into phases: an inactive phase, known as *anestrus;* a time when the eggs are growing, called *pro-estrus;* then *estrus* itself, when the egg follicle ruptures, releasing the egg (in many species this seems to stimulate overt sexual aggressiveness); and finally *met estrus,*

A copulating pair of cockroaches (Xestoblatta hamata).

when the egg has moved out of the ovarian area into the oviduct. When this happens the female apparently loses all interest in copulation. During the third phase, the peak of estrus, a mare that may have ignored all the stallions in her neighborhood for the past thirty days moves aggressively toward them, rubs against one male in particular, and agitates for copulation. The Greeks, who first observed that some female animals seemed driven to distraction at this time, attributed their behavior to the presence of a "gadfly." Estrus means quite literally that the mare, baboon, or dog is driven to distraction as if a fly were trapped inside her, pushing her to strange behavior.

A female chimpanzee in estrus at Gombe Stream Research Center.

"Driven to distraction" may be an exaggeration, yet it is true that animals in estrus hawk their condition. Many display a visible sign of sexual readiness, such as the pink, swollen sexual skin on the female chimpanzee or the crimson swelling of the mandrill. The visual signal advertises from a distance that these females are in the midst of producing an egg and are ready to be fertilized.

Estrous and menstrual cycles are really expressions of the same process, with emphasis at a different stage. There is no apparent evolutionary connection between the vivid sexual swellings, menstrual synchrony, and the taxonomic charts. Chimpanzees have a bright estrous skin, but their close relatives the gorillas and orangutans do not. The small, monogamous New World marmosets do not have a noticeable estrus, while middlesize African primates like baboons do.

What actually triggers ovulation differs from species to species. Ultimately the maturation of the ovaries depends upon the female's endocrine system. In insects the trigger seems to be the corpora allata, a tiny pair of glands behind the brain. In vertebrates the pituitary gland at the base of the brain starts a series of hormonal cycles that periodically render the female fertile. Follicle stimulating hormone (FSH) causes the ovaries to mature and begin to function at puberty. This hormone also stimulated the ovary to produce estrogen. A second pituitary product, luteinizing hormone (LH), cooperates with FSH to bring about ovulation. Luteinizing hormone causes the ruptured follicle to become the corpus luteum and secrete progesterone, which causes the uterine lining to thicken so it can receive the developing embryo.

The estrous cycle is governed by the female's maturing process as well as the climate and length of day. It is curious that animals remote from mammals, such as the common cockroach, produce eggs on a monthly cycle. In these females, ovulation is spontaneous; whether or not a male is around to fertilize the eggs, the female produces them on schedule. But a small number of mammals function differently. Cats and domesticated rabbits must copulate first in order for ovulation to take place. The egg does not enter the fallopian tubes until triggered by either the movement of copulation or perhaps by male hormones expelled with the sperm. Laboratory studies show that cats must copulate at least three times before ovulation can occur, which probably accounts for the exceptionally sensual postcoital actions of most cats. After intercourse, the female continues to try to arouse her partner, and can wear out more than one male as she encourages copulation until she ovulates and becomes pregnant. Although postcopulatory ovulation has been directly observed only in a few mammals, there is some evidence that female human beings will ovulate under the stimulus of copulation.

Copulation also stimulates ovulation in colonial birds such as the gull, which responds to the sight or sound or smell of other gulls that are performing sexually. The synchronous ovulations of bird populations was first noticed by Fraser Darling in 1939. Observing the breeding habits of herring gulls, a colonial species that lives in great numbers off the English coast, he discovered that the first bird to breed set off the others, and that the earlier the breeding, and the greater the synchrony among the gulls, the greater the success in terms of live births. He also noted that the larger the size of the colony, the earlier the breeding and synchrony. In laboratory tests to explore this phenomenon, he discovered that merely introducing a third bird into a cage of breeding parakeets stimulated the caged female to ovulate. Synchronous ovulation occurs throughout the animal kingdom and it is still not altogether understood. Pheromones as well as some psychological factors are probably involved.

Seasonality

Animals that have "seasons" do not ovulate year-round. Most fish, amphibians, reptiles, and birds fall into this category, but so do enough mammals that seasonality cannot be said to be exclusive to any particular order. The ovaries in females of these species remain dormant for a great part of the year. Then, as the temperature changes and the days lengthen, the eggs begin to acquire yolk and the female carrying them is ready to be fertilized. After fertilization, she will begin laying eggs or carrying an embryo inside of her.

Seasonal animals vary as to when they ovulate, and for how many cycles. Many mammal females go into estrus only twice a year, often in summer and winter. Those in temperate zones experience seasons, while those near the equator merely know a change of precipitation. Denizens of temperate climates respond to longer daylight hours and rises in temperature in different ways. Some escape the rigors of winter altogether and enter a state of hibernation; their body functions slow down, conserving energy until warm weather returns. Hibernating females emerge in the spring, and some, like the brown bat, release fertilized zygotes (another kind of suspended animation) which proceed to develop as embryos. Others, like swifts and redpoles, begin to cycle as soon as they awaken and quickly find mates. And metamorphosizing insects like caterpillars react to the change of seasons by weaving cocoons and emerging much later as moths or butterflies.

Birds in temperate zones cycle for only a small part of the year. The longbilled marsh wren *(Telmatodytes palustris)*, for instance, copulates many times a day during this short fertile period, and will lay half a dozen eggs before she settles down to brood them. She may produce several families before she stops ovulating for the year.

Sheep in the British Isles begin to ovulate as the days grow shorter at the start of winter, become pregnant then, and bear their young in the spring. Yet English sheep that have been brought to Australia and New Zealand are fertile all year long. Other animals that have traveled from the northern to the southern hemispheres have altered their cycles to conform to the changed position of the sun, suggesting that the reproductive cycles are photosensitive, triggered by the intensity and length of daylight. Species such as macaques that mate seasonally in their native Asia cycle continuously in North American laboratories when the light is held constant, which seems to confirm the significance of light. Yet animals such as squirrel monkeys *(Saimiri sciurrus)* follow an inner clock and go into estrus at a certain time each year whether they are in a tropical rain forest or in New England.

In most animals the male undergoes physiological changes that parallel those of the female. Among species such as giraffes and zebras, males may leave the masculine company they have been keeping most of the year and enter into a stage very much like estrus in a female. Rutting males behave as if they too have a "a buzzing in their ears" as they pace and snort and sniff at females, doing everything they can to copulate as often as possible.

Some females have no apparent cycle at all, seasonal or otherwise. That is to say, they ovulate only once and risk their whole chance at genetic immortality in what some geneticists refer to as the "big bang." This strategy occurs in animals that live just a few days, as well as others that wait a decade to reproduce. The tiny aphid-egg mite *(Adactylidium)* allows her offspring to eat her own innards as they develop inside her body, ultimately destroying their mother in her single act of reproduction.

Among females that are sexually receptive for only a limited time, some, like the mosquito, accept a single mate, so that the one copulation—and one father—has to do the trick. Others, like the Belding's ground squirrels *(Spermophilus beldingi)* that live in the lush meadows near California's Yosemite valley, though in estrus only four to five hours in a year, mate with from one to eight males during that time. If the female survives the rigors of an alpine winter, she can have another few hours of sexual encounter twelve months hence. Fur seals (Phocidae) in the Kurile Islands off Japan, which are in estrus for a single day each year, appear to mate with as many males as possible. Those females that copulate repeatedly during their short spell of fertility accept the spermatozoa of three or four animals. Belding's ground squirrel litters are usually multiply sired, as are those of deermice and various members of the cat family. Among honeybees the queen bee accepts the sperm of—or, in the parlance of entomologists, "fecondates" with—a dozen drones in the few hours of her annual nuptial flight. Later, when back in her hive, the queen bee releases what sperm she will at the appropriate time.

Animals that live only a season often spend a disporportionate amount of their lives in copulation. This is especially true of butterflies, many of which, like the checkered white female, mate for half an hour at a time, every three or four days, throughout the two-week period that is all they know of this world. In contrast, many long-lived animals spend a much smaller proportion of their adult lives copulating. Among some that live for at least a decade, the female ovulates only once a season, with her fertile period so timed that her young will be born when the temperature is mild and food is abundant.

Others copulate even less frequently. The blue whale *(Baloenoptera musculus)* carries her developing young within her for more than a year. After its birth, she nurses it on fat-rich milk for at least six months. Females are seldom ready to mate again before two years have passed since their last encounter, and more frequently the interval is three years. Because of their deep-water environment, whale matings have not been closely observed, and the estimates of copulations are based on a pattern of births and observations of the rare captives.

As far back as Aristotle, observers have noted that elephants, the largest land animals, are seldom pregnant. Why this is so is still not understood, but new biochemical studies of African elephants suggest that females produce very small quantities of progesterone, the hormone manufactured in the corpus luteum that maintains pregnancies. The elephant has to accumulate a supply over several cycles in order for a fertilized ovum to become successfully implanted in her uterus. When that occurs, the female will carry the fetus for a year and a half, and then nurse if for five years or more. She is only ready for the cumbersome task of sexual intercourse once every four years, and then for only a short time, according to the observations of the ethologists Ian Douglas-Hamilton and Holly Dublin.

Size alone does not account for long interludes between copulations. It has been suggested, though never proved, that reptiles like the common prairie rattlesnake follow a two-year cycle in which the female sometimes holds the sperm throughout the winter before they fertilize her eggs in spring. The eggs then develop ovoviviparously inside her body for another eighteen months, after which the young are born alive. Female green sea

African elephants copulating in the water.

turtles in remote stretches of the South Pacific copulate every few years at sea. Then they come ashore about once every eighteen months to nest and lay eggs, and later return to bury their eggs in the sand.

A Time and a Place

Like many insects, salmon (Salmonidae) and eels (Anguillidae) spawn only once in a lifetime. For these animals, the place is just as important as the timing. The habits of the freshwater species of eel had mystified ichthyologists until Johannes Schmidt spent the eighteen years before 1922 searching for their spawning grounds. He found lines of equal-sized larvae in a circular pattern centering on the Sargasso Sea between the West Indies and Bermuda. He deduced that eels spawn there, then die, leaving their newly hatched transparent offspring to make their way in a year's journey to the rivers of Greenland, North America, or Brazil, or a three-year swim to the coasts of Europe.

Many species of colonial birds seek out special breeding grounds where, for a few weeks or months a year, they ovulate and copulate. Arctic terns *(Sterna paradisaea)* mate on the frozen tundra, where they remain until the hatchlings can fly with them in their long migration across the globe to Antarctica. Some water birds, such as New Zealand's mutton birds, are reputed to avoid all land for the months they are not brooding. Later they mate, then leave for a month-long foraging trip before they settle down on small rocky islands or reefs. Here they dig and line holes in the ground as nests for their expected offspring.

Less dramatic are the ways of the swallows that faithfully return each March to the region around the old mission at Capistrano, California. Among mammals, pinnipeds such as the gray seal *(Halichoerus grypus)* return to the same rocky beaches annually, just in time to deliver a pup. They copulate immediately after delivery, in what is called postpartum estrus, and depart for the sea as soon as the new pup can fend for itself.

The timing of copulation is crucial to the successful arrival of a new generation. Offspring conceived at the optimum moment will be born when the climate is right and the most food available. But timing is not everything. Where the females go to conceive is also crucial, as are the partners they select.

seven

The Company She Keeps

DARWIN recognized that the overwhelming advantage of sexual reproduction is the flexibility it provides when environmental change threatens the species. Most recent biologists agree that sex makes sense, because the chances of successive generations adapting to new conditions are better when half the genetic contribution to the offspring comes from another member of the species. Yet which male a female mates with, and how many partners she has to choose from, make the apparently simple exchange of pieces of chromosomes a complicated matter.

For a long time it seemed to zoologists that living systems and mating systems were the same. However, recent studies with elephant seals indicate that the large bull, ostensibly responsible for inseminating all the cows, relinquishes some of them to younger bulls as the season draws to an end. Likewise among birds that appear to be monogamous, blood samples of offspring indicate that there is often another father involved in insemination, though perhaps not in the contribution to brood care.

The living and mating arrangements among animals exhibit a broad spectrum. Classic *polygyny* finds a single male herding lots of females in a harem. In mammal harems, the male leader is frequently a great deal larger than the females. The extreme sexual dimorphism of the male gorilla, with his barrel chest, represents one pattern. The infant apes look so much alike that zoo-keepers have waited months to know which sex has joined their group, but by maturity they have altered so much that the male is 25 percent larger than the female, reaching about 250 pounds to her 200.

His head develops an impressive crest, reminiscent of a roman helmet, and the saddle area of his back turns silver. Leading his troop around the slopes of the dormant volcanoes in the east African highlands, the male mountain gorilla *(Gorilla gorilla beringei)* stays in a staked-out territory unless the food supply should for some reason fail. During foraging expeditions, he devours the same leaves and wild celery that the females and offspring do. The silverback's entire following—females, infants, sons, and brothers—all look to him for protection.

The male leader of a troop is ready also to defend his position against the lone males who occasionally try to take over. The rival rears up on his legs and drums against his chest until the resident male responds. Shaking the branches on nearby trees, they send leaves flying and fill the air with a

A male mountain gorilla displaying for the benefit of anther male.

pungent body odor. If their own leader has grown old or become ill and loses to the intruder, most females change allegiances. The female's loyalty is to her own well-being, and by inference to the well-being of her living young and those she is yet to bear.

A possible explanation of the connection between polygyny and sexual dimorphism comes from Timothy Clutton-Brock, a British ethologist whose research team is studying red deer on the Isle of Rhum in Scotland. Here the large stags have a short breeding life, and the smaller hinds are reproducing mothers for many years. Because there is such a contrast in breeding strategies between the sexes, he suggests that males need greater size early in life to outmatch other males if they are to breed at all. The females' longevity enables them to bide their time.

Polygynous systems do not always show such a vast disproportion in the number of females and males. Although almost 95 percent of mammal females share males, there are more often just two or three females with a larger, but not gigantically larger, male. Whereas some harems include vast numbers of herded, guarded females, smaller polygynous groups may be longer lasting and have a more rigorous social structure. The dominant female thus receives the bulk of the male's attentions, such as more food and a stronger nest. The others live off the leftovers, moving up in the hierarchy, perhaps, if the favored female vanishes. Female ostriches *(Struthio camelus)*, for instance, follow a male at the start of the courting season, until they join a group of other females at his breeding site. There they wait. One at a time they speed off to a food-pecking rendezvous with the male (a ritual that always precedes copulation). But the dominant female is usually the first to copulate. She and the male will incubate all the eggs, thus determining which eggs will be tossed and which allowed to hatch. She behaves as if the first sperm is more potent, as do females of other species such as fruitflies *(Drosophila melanogaster)*. These females are wary of depleted sperm and actively seek out virgin males. Females in species such as fruitflies are called "promiscuous." They court many males and do not seem to establish any kind of bonds. Unlike fruitflies, however, most females conform to some social pattern, if only for a single breeding season, and confine their advances to one or a few males of their species.

Classic *polyandry*, often referred to as "role-reversal," with one female dominating several males, has been observed by Donald Jenni among the marsh-dwelling jacana. In the water meadows of Costa Rica to northern Mexico, the female bird dominates her territory, chasing off intruding females, then swooping down low over each of her mates to entice him into aerial acrobatics. About three-quarters again the size of the males, the females divide the territory between them and include the males in the same way that serfs went with the territory in prerevolutionary Russia. The males, if they choose rather than allow themselves to be chosen, probably go with the female who controls the best territory. Some females are left out of this pond version of musical chairs at first. But as the rainy season, which is the breeding season, progresses, the water rises and produces increasingly more marsh sites, so that eventually most females get a chance to breed with as many mates as they can sequester.

Most species in which females outsize males live in assorted systems of polyandry. Each female courts and keeps at least two but often three or more males. Greater rhea females *(Rhea americana)* clan together on the broad pampas of northern Argentina. Here they allow a single male to herd up to fifteen of them together in what looks at first like a harem. The male builds a sort of communal nest, to which each female returns several times, leaving him with as many as fifty eggs to guard. But after a while the females wander away and allow another male to "herd" them. They lay eggs for him as well. Selecting as many as seven males a season, female rheas leave their eggs to be guarded and hatched by each of their seven suitors. Ornithologists call this behavior *sequential polyandry,* because the female moves from mate to mate but does not consort with more than one at a time.

A different situation confronts the Tasmanian hen *(Tribonyx mortierii),* a native of Australia. She is courted by a pair of brothers, and lives with both simultaneously, laying eggs in a single nest. A study of one population by the British population geneticist John Maynard Smith revealed that both brothers were seen to benefit by this arrangement. Although one of the males usually dominates his brother and obtains two-thirds of the copulations, all three shared incubating and nesting care. Maynard Smith suggests that for some reason there is a shortage of females in this species, which makes it advantageous for brothers to cooperate, by sharing a mate and a territory.

Some owl monkeys in the Amazon forest also live in polyandrous groups. Although there are some mating pairs, groups with a single reproducing female and two or three males have far greater reproductive success. The males carry the infants in this species, and as they are sought after by many predators, it may be that two males are necessary to ensure the survival of at least one youngster.

Many species live *monogamously* in stable pairs, and they often reflect the observation that members of couples that live together look alike. Animals that remain paired for more than one season tend to be sexually monomorphic; that is, they are about the same size and weight and are difficult for observers to distinguish. Monomorphic species live all over the world, and are not limited to any family, appearing in fish, birds, and mammals. Female and male siamangs, small Asian primates, look alike and weigh about the same. Their intimate relationship throughout a lifetime is remarkable in its steadiness. Likewise, tiny marmosets in the forests of Central and South America are also primates of equal size that live in pairs for long periods of time. Male and female Amazon parrots are monogamous and monomorphic, as are small butterfly fish in the Red Sea.

Monogamous species, even such short-lived ones as the tropical angelfish, indulge in extended courtships. It is as though they sense the proportion of their lives they will have to spend together and are thus especially careful as they check each other out. Among highly complex rodents, beaver couples often choose each other as adolescents and live together for at least six months before they are sexually mature enough to take courtship to its logical conclusion. What the males and females of each species of these monogamous animals have in common, besides the same general body size and shape, is a joint interest in defending their territory from other males of their species and sometimes from their own progeny as well.

Monogamy is also found among charedonts, marine fish that live in stable pairs for years on end. Neither female nor male cares for the offspring. Their permanent bond may have evolved from mate guarding, an effort to ensure that no other male has access to the female, and the female's parallel interest that the male does not spread his sperm elsewhere. This long-term heterosexual bond is also found among soldier beetles, where the apparent devotion of the pair to one another appears to be the best strategy for both sexes to ensure continued access to a mate.

But all monogamous species are not monomorphic. Perhaps the ultimate monogamy is found among the angler fish, which inhabit the deepest ocean waters. As fry, female and male are almost the same size. They are free-swimming and have enough yolk to forgo eating, which is fortunate as males have no digestive system. They must find a mature female who is a great deal larger. Fully grown, she attracts a male, who bites her, permanently—never releasing his hold. She continues eating and swimming with her mate attached to her by his mouth. Gradually their circulatory systems join and the female provides the male with food and oxygen. When she spills thousands of eggs into the water, the living sperm bomb she is supporting ejects sperm to fertilize them. But by the time this

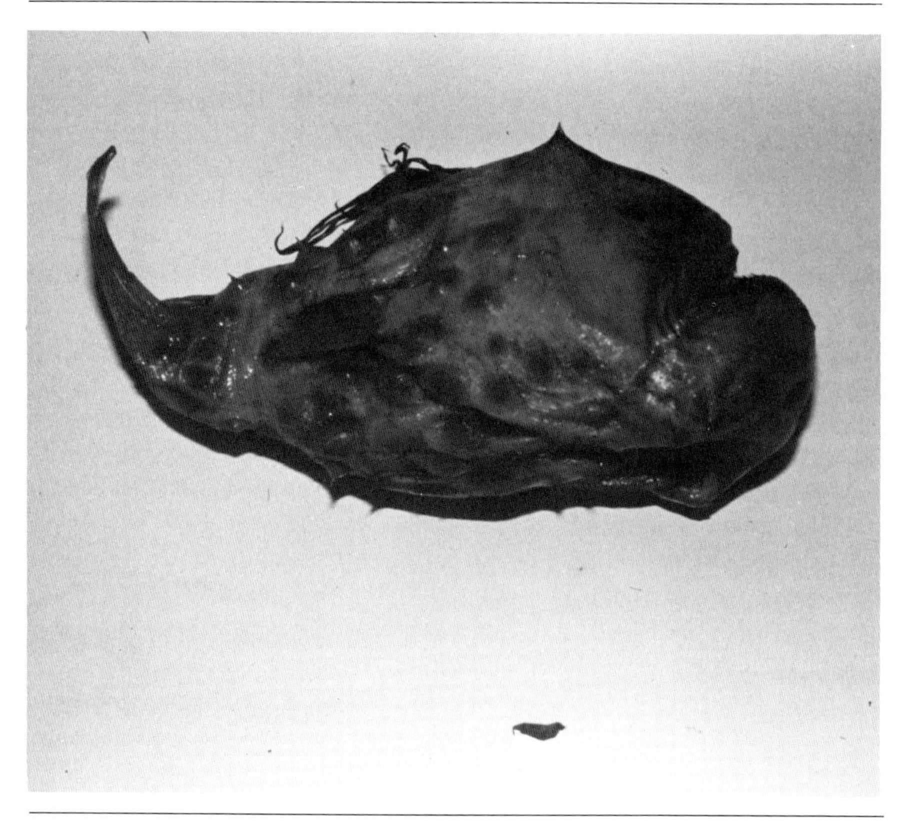

Acute sexual dimorphism in a species of angler fish. The female is on top; the male is beneath her.

happens, she will have grown to several thousand times his size, and he will remain a minute appendage on her body.

Monogamy creates stress, which may account for its relative rareness among longer-lived and larger-brained species. Monogamous animals must tolerate one another, even when they are not courting or raising young. In some species the members of the monogamous pair divide responsibilities along sexual lines, such as the owl monkeys whose males do much of the infant rearing and whose females concentrate on finding food. But in other species, except for the obvious job of producing offspring, they blend roles. Females of the species usually care for the young and let the males help guard them from danger and help procure food. The female beaver helps build the lodge, but once she has settled in and borne a litter, she remains inside and lets the male keep it in repair. And some carnivores team up to guard their territories, female and male alternating urine-marking to warn away intruders.

Birds of prey live in mated pairs in which the females in species like the sparrow hawk and eagle weigh one third more than the males. The explanation for the extreme divergence in size is open to conjecture, but seems to reside in how they get what they eat, rather than how they court each other. Some flesh-eating species, such as vultures, feed entirely off carrion and are thus the least dimorphic of the lot. They can pick up as much flesh as they want without a struggle. But species that catch their meals alive, on the wing, tend to have the largest females. While she broods the chicks with a body large enough to cover every one, her mate hunts, his smaller size allowing him enough speed to capture fast-flying prey, which he brings back to the nest. But though the male's smaller size may have evolved as an aid to procuring food, the result in terms of behavior shows that the larger female uses her greater size apparently to tyrannize her helpmate—she pecks at him, pushes him away from the very food he has brought back, and even keeps him away from the nest. Whatever the adaptive value of diminutive males in terms of providing food, their size results in what looks to an observer like victimization. Larger females act with blatant aggressiveness toward males and plainly dominate them in their daily encounters over food or when simply crossing one another's paths.

Many monogamous animals are monomorphic, but monomorphic animals are not necessarily monogamous. Among the spotted hyenas in the

Serengeti, the sexes seem identical until the female has given birth. Once she becomes a mother, her nipples stand out, making her sex obvious to an observer. Adding to the confusion between hyenas, females of this species, through an apparent overabundance of male hormones, have developed pseudo-penises and scrota. The similarity between the young of both sexes is so great that many observers find it impossible to distinguish the young without a close physical examination. Yet spotted hyenas live in large, female-dominated groups; they are not monogamous. Likewise, the young mountain sheep of North America also look alike, the females growing horns as do the males, whose greater horn development continues after sexual maturity. These young females often behave like males, even exhibiting what is called "rutting behavior." Like sexually excited males, the young females butt each other with their heavy horns and mount each other in what looks like homosexual play. Yet female and male mountain sheep never live in pairs. Rather, outside of the spring mating season they graze about in large sexually segregated herds.

Evidence from the cases of hawks and adult hyenas suggests that, in mammals and birds of prey, the sex which rears the offspring is often larger, providing both more protection and warmth. In 1976 Katherine Ralls reviewed the disparate literature about size dimorphism in mammals. She suggested that bigger mothers are better mothers. Female animals as well as birds of prey use their larger size to protect and provide more food for their young.

Females are larger than males in enough species of birds and mammals to make the phenomenon more than an idiosyncratic exception. Females outsize males in most species of bats, in several whales, and often among the varieties of New World monkeys. Among invertebrates and many reptiles, amphibians, fish, and birds, females are frequently larger than their mates. Throughout the animal kingdom, females outsize males in at least as many species as those in which they do not. And though exceptions abound, there is a distinct relationship between the relative size of the sexes and the species' social system.

Paired couples, large harems, and small male- or female-dominated polygynous groups are among the most frequently observed social units in which most social animals choose to live. Yet other arrangements can be found. Females with dependent young may live in social groups that do not include males. For some species, such as lions, females outnumber the males most of the time. For others, such as chimpanzees, there seem to be almost equal numbers of both sexes. Another pattern finds females like elephant seals living alone much of the year and congregating in pods for the few months they come ashore to bear their young and become pregnant again.

But a few species seem to be true loners. No mammal mother can be completely solitary, but the cheetah *(Acinonyx jubatus)* tries. When ready to mate, the female solicits a whole harem of males from among whom she makes her choice. Gathering as large a following as she can, this feline initiates a chase, leading a band of four or five lusty males for hundreds of yards in a straight line across the plain before she selects one from among them with whom to mate. After this she is alone until the birth of the cubs that she relinquishes after only a few months, and then she is solitary again.

eight

In a Pumpkin Shell

THE EVOLUTION of sexual reproduction and female choice of mates has not conferred upon females unlimited control of the reproductive process. Evolutionary success is a history of compromises between both males and females, and between a species and its changing environment. The meshing of reproductive strategies—from anatomy to behavior—is crucial to a species' survival. Often the self-interests of the sexes are at odds. Successful species achieve a balance of sorts, but seldom as secure as either sex might prefer.

Promiscuity in a female does not suit most males, for good reason. It is in the male's interest to have some assurance that his sperm will be used in creating the next generation. Thus males of many species have devised mechanisms to protect the success of their copulations. Although some of these arrangements seem unduly complicated, they apparently suit the females of the species, since they have acquiesced. These females have, in effect, traded off a guarantee of safety and sustenance for their offspring in exchange for an end to "shopping around" and freedom.

The female hornbill, a fifteen-inch-high monogamous African bird whose bizarre bill and plumage easily distinguish her from other birds in Ethiopia, cooperates with her mate in completing her own sequestration. The hornbill couple scan their territory together and select a place to build a nest in the hole of a dead or dying tree. After copulation, the female enters the hole and begins to seal it up from inside with her droppings. The male assists by bringing her bits of mud. They close the hole until only a small opening remains, through which the male feeds the female for the five weeks during which she lays her clutch, an egg a day for a week. After the chicks hatch, the male continues to feed his family until they are ready to leave. Then mother and chicks break down the barrier and fly away. The trade-off is clear: her voluntary incarceration results in a high proportion of successful hatchings. In fact, what could disastrously limit her breeding success would be the untimely death of her mate.

Female mosquitoes emerge from the larval stage exuding a powerful

A male hornbill feeds his sequestered mate through a hole in a hollow log.

pheromone that attracts a suitor. The first male to arrive inserts his intromittent organ into her spermatheca immediately. But along with his spermatophore, the male mosquito deposits a chemical called a "matrone" that more than counteracts the pheromone that attracted his attentions in the first place. Its offensive odor makes the once-attractive female suddenly repulsive to all other males. Left with a supply of sperm, the female mosquito deposits her eggs one by one, releasing sperm for each egg until she has used up her supply, and her short life ends. Likewise, the female butterfly *Heliconius erato* accepts the male's "stink odor" (a smell like witchhazel) that repels all other males, discouraging them from copulating with her, and is thus limited to a single supply of sperm. She must somehow benefit from this transfer, for she has developed a special ventral sac to receive the fluid.

A more common form of female accommodation to male pressure is a female's physical acceptance of a vaginal plug, putting a cap, literally, on the mating procedure. Mating plugs vary enormously from species to species, but their function is the same. The male *Johnnesisenlla nitipta* fly leaves his genitals as a stopper to make sure that no other male can ever inseminate his mate, and in this suicidal thrust he provides a protein-rich meal for his offspring. Other males, like baboons, shrews, and spiders, merely leave a temporary plug. The male seals in his own sperm for the lifetime of the sperm and often for the fertile season of the female. A plug is a male's response to a promiscuous mate.

Although some female spiders have both pouches of their epigynums filled with sperm after a single copulation, others do not and copulate with a second male to fill them. Some continue copulating after the eggs are laid. It is unknown why these female spiders solicit copulations after they have laid their eggs, because these sperm are then useless. But the male cannot be aware of that. Perhaps in a gesture to prevent the sperm from leaking out, or mixing with a successor's donation, male spiders of the comb-web species place a hard conical cover in the female's epigynum that remains permanently and seems to become part of her anatomy. Plugged up like this, the female guards the sperm, and interested males do not waste their efforts on her but find another female whose ducts are open.

Mammals, too, accept vaginal plugs, including such diverse animals as rodents, shrews, opossums, baboons, and some hooved females such as the collared peccary, a close relative of the familiar barnyard pig. Peccaries live in herds in scrub lands and swamps all over the New World. They are sexually playful all year round, females mounting females regardless of the estrous cycles. When they do come into estrus, the females court several males. Each male leaves a copulatory plug, a solid object weighing about 30 grams, along with his sperm. The plug acts as a barrier between his semen, which is safe behind it, and new sperm from a competitor.

Sperm plugs in peccaries have not been examined as thoroughly as those in reptile species, which are more easily studied in laboratories. Herpetologists have discovered thick gelantinous plugs in the cloacae of garter snakes and deduced that although they do not remain in place long enough to guarantee that a male will be the sole father of the female's brood, they do discourage other males from copulating while they are in place. Copulation in garter snakes usually happens six to eight weeks before the female ovulates. Of course any female could copulate again after she has managed to expel the plug, but most wander away from the mating territory before that time. While the plug is in place, the sperm can reach specialized storage pouches in her oviduct walls so that when she finally produces eggs, the sperm she received earlier are in place to fertilize them.

Copulatory plugs most certainly reduce the female's efforts at remating, but they do not, in most species, prevent it altogether. *Parnassius* butterflies have been found with more than one plug, and female ticks have been found with as many as five spermatophores inside them. In these cases the extra plugs help the female in more than one way. In addition to giving her a choice of sperm, she gets the additional nutrients each plug provides.

Plugs play a double role. The first is keeping the sperm safely near the ovaries until the eggs are ripe, thus helping the female achieve her goal of fertilization. The second aids only the male, for it prevents the female from seeking another mate. Those females that accept plugs seem to have evolved according to the rationale that it is a safer bet to retain the supplies of sperm they have rather than to keep copulating in the hope of finding fitter fathers for their offspring.

Nature appears to have an infinite variety of ways to adapt a contradictory strategy to meet new situations. Mating plugs may help some females and males, but there is at least one kind of damselfly that has successfully found a way around them. These insects live in great abundance on the shores and waters of New England rivers. Unique, as far as has been ascertained, is the male's dual-function penis. Like other males, he uses this organ to introduce sperm into the female. The female's spermatheca is usually filled after a single copulation with one male, but that does not deter a second male of this species. His penis is equipped with a pump-like mechanism which empties out the sperm deposited by his predecessor as he fertilizes anew the female of his species. Equally curious is the behavior of the fertilized females. They actively seek new mates, allow their original supply of semen to be removed, and accept a new supply, often allowing that to be refilled several times until a wary male takes it upon himself to guard the female and hovers near her until she finally oviposits, using his semen alone to fertilize the eggs. How these multiple copulations benefit the female is a mystery, unless by some unknown mechanism the female is able to judge the male's fitness by the stimuli she receives during copulation.

In some species, even successful fertilization by one male does not prevent another male from turning the situation to his advantage. In 1959, in an experiment with laboratory mice conducted by Hilda Bruce, a female mouse about forty-eight hours pregnant suddenly resorbed the embryo before it had become successfully implanted in her uterus, just after a strange male mouse was introduced into her cage. At first glance, and indeed for many years afterward, this phenomenon (called the "Bruce effect") seemed to suggest that female mice, in the interests of heterogeneity, prefer strange, rather than familiar, fathers for their offspring. Since then, embryonic resorption has been attributed to the displacement of one male by a more dominant individual. It has been reported in a number of species.

Recent research by E. B. Keverne and C. de la Riva reveals that the male pheromone that prompts the resorption is the same odor that earlier in the female's life triggered puberty and first estrus. That is, the same male pheromone suppressed the female's production of prolactin so that she began to cycle. However, prolactin plays a different role in the female cycle after fertilization from the one it plays at the start. Its sudden suppression at this stage prevents successful implantation of the fertilized egg. Among mice, the act of copulation is accompanied by an imprinting on the female of her own mate's pheromone so that her prolactin production is not affected until the fertilized egg is safely in place. But the sudden appearance of a strange male with a different odor prompts the suppression of the hormone. This event probably would not occur among mice in the wild, but among other species the appearance of a strange male is frequent, and the resorptions that follow the displacement of one male by another may be caused by this hormonal mechanism.

U. W. Huck reports that dominant males block pregnancy more effectively than their social subordinates, perhaps owing to a greater intensity of the male pheromone in these individuals. Whatever the cause, and whether the chemical mechanism is identical in all species, the phenomenon is not unique. Preimplantation block (resorption) has been seen in wild mice, collared lemmings, and voles. Postimplantation block (abortion) has been reported in several species of voles and deermice, while midgestation block has been reported in one species of deermouse. It seems to be yet another reproductive compromise between the male and female in some species: she trades away certain fertilization with the sperm of a previous male for the opportunity to start the process all over with a seemingly better chance that her offspring will survive.

That females in many species seek multiple partners cannot be doubted, but the advantages (beyond immediate sensual gratification) are not always apparent. A possible evolutionary explanation of promiscuity in species like Belding's ground squirrels, which produce litters with multiple fathers, is an extension of the rationale for sexual reproduction in the first place. If one father provides chromosomal variation, then five fathers for one litter will provide five times the variation and thus increase the female's odds that some of her offspring will be able to cope with an unpredictable future. For females that only produce a single pup a season, such as elephant seals, mating with several young males in addition to the great harem master provides a different kind of insurance. The harem master copulates almost without resting as, one after another, the females he has gathered together come into estrus. It is quite possible that after a few days he is literally worn out, his sperm count severely diminished. It is at this stage of their annual adventure that female seals allow the young males who have been lurking on the fringes of the group to interrupt their copulations so they can copulate with them before they seek the water again for another long winter.

Female chimpanzees alternate between promiscuous and monogamous relations when they enter their monthly estrous period. At the start of estrus the female seems to copulate with almost any available male, regardless of rank or age. But as she reaches the peak of her sexual swelling, she often disappears with a single partner. Then the pair avoid the rest of the troop and copulate frequently throughout a two- or three-day interval. Her promiscuous behavior may have no other evolutionary purpose than to provide ties with the other, often younger males in the group, so that they will not harm her offspring. But an equally reasonable, and less remote, possibility is that the earlier copulations were necessary to stimulate the production of hormones so that the egg, when fertilized by the dominant male, will implant in her uterus.

A female chimpanzee at Gombe copulates with a second partner.

The frequency with which a female animal seeks copulation may depend on how often she needs fresh sperm. Among rattlesnakes *(Crotalis)* throughout the Americas, females can keep sperm alive for many months from a single copulation, even years in other species of snake. This single fertilization, plus the added time that it takes for a snake embryo to develop, assures the female that she need not copulate again for several years. Likewise, some female marsupials such as Australia's well-known red kangaroos cycle differently from most other mammals. The female kangaroo has a split or double uterus, so that when one part holds a developing embryo, the other continues under hormonal influences. Thus she continues cycling throughout any pregnancy and is able to allow one egg to develop into an embryo while having another egg fertilized. Some insects like bees and butterflies store live sperm for days, and sometimes weeks and months.

Some sperm, however, are viable for only a few hours. Dogs, both wild and domestic, meet the problem of short-lived sperm by the copulatory tie that literally holds female and male together until the sperm have reached the eggs. The sperm either succeed or fail. If they reach the eggs, they fertilize all of them at once, ensuring that canine litters are always full-siblings. This is in contrast to the cat family, where the female can count on sperm living for as long as ten days. She continues mating with new partners, often giving birth to a litter of half-brothers and half-sisters. In the case of felines, it has obviously been advantageous to a female to produce a mixed set of offspring. One or two individuals may indeed be more successful than the rest.

Barbary macaques—large terrestrial monkeys that live in groups in the cedar forests of Morocco—do not produce multiply-sired litters, but their promiscuous behavior does seem to have other reproductive advantages. Female macaques, like female chimpanzees and baboons, advertise their sexual readiness with a swollen rump, and they engage in a series of copulations with many males. Interested females seem unconcerned about the marked rank among male macaques. They approach and solicit the sexual attentions of one male after the other, copulating with each one in turn. In effect, the female eliminates any demonstration of favoritism by favoring all the males, and in the process offering all of them some probability of siring her offspring. David Taub has hypothesized that the attentions of not one but several males, any of which could be the father, toward a newborn macaque make a crucial difference in the infant's chances of survival. The promiscuous behavior of the female, then, ensures not only fertilization but critical help with the rearing of the offspring she will eventually bear.

Incest

Sequestration, vaginal plugs, and spontaneous abortion are not the only methods species use to control reproductive behavior. Where incest is a potential problem, a number of mechanisms have evolved to limit the practice. Sexual reproduction has its pitfalls, of which the most likely is the expression of traits in the new generation that will weaken the offspring. Humans in most cultures have long recognized the danger of inbreeding —mixing the genetic contribution of brothers and sisters, fathers and daughters, mothers and sons. Humans are so familiar with the taboo against incest that some biologists, as well as anthropologists, assume the aversion is innate.

There is also evidence that complete outbreeding—mating between two animals with little common genetic material—may be disadvantageous. Artificially-bred individuals such as female horses and male jackasses usually produce sterile mules. In nature such inter-specific exchanges possibly produce the same result, sterility. There is probably an optimal degree of relationship within some species, a balance between too close a familial connection and too remote a connection, that serves to enhance the successful continuity of the species. To strike the right balance, an animal needs to discern those individuals that are close kin from those that are not. The mechanisms that allow an individual within a species to detect relationships are slowly becoming apparent. This is a hotly disputed topic just beginning to be understood as evidence in species from insects to mammals reveals that animals recognize degrees of kinship, and act toward related individuals thus recognized in ways peculiar to each species.

Within the animal world there are communities where incest is the only option. Moreover, these communities are ancient, which suggests that the practice has not caused any apparent degeneration of the species. The tiny *Adactylidium* mite lives in parasitic harmony with the larger thrip. Immediately after her birth, she attaches herself to a single thrip egg and soon produces between five and eight daughters and a single son. This fellow plays the double role of brother and husband to his sisters. The mite family lives out the greater part of its communal childhood in the safety of the mother's body, feeding on her tissue. The offspring are still inside her when the now-mature male copulates with all his sisters. Eventually the young devour the rest of their mother, killing her, and break through her body wall in search of more food. At this point the females are all pregnant. The male dies, his mission accomplished. The females find themselves more thrip's eggs and continue their incestuous life cycle. Since they always copulate with a brother, the genetic variation possible within such a system is smaller than it is in other species, although there is still variation. Recombination has evidently been sufficient to allow the species to survive over evolutionary time.

Insects are not alone in their incestuous unions. Studies of fallow deer *(Dama dama)* in Britain reveal a pattern of incest among these graceful ungulates. Protected in a royal preserve, fallow deer have coats of varying colors which have made it easy for forestry commission rangers to keep track of and report on the habits of individuals within the herds. The deer move about in distinct groups, the adult males together, and the does with their fawns. As the days grow shorter at the end of October, each great buck returns to his rutting stand, the same territory he has returned to for three years, where the same doe seeks him out annually. Until a younger buck replaces him, the doe will copulate with the same buck year after year. Important to the future of her species, she brings along any fawns she may have had. By the second year a female fawn is sexually mature and will copulate with her own father, to whose side her mother has brought her. Although this is very close inbreeding, the rangers who have kept watch over the deer population thus far report no apparent ill effects. It seems that ongoing inbreeding has not harmed this species, and may in fact be in some way selectively advantageous for it. This kind of inbreeding seems to be limited, in fallow deer, to fathers and daughters. It rarely occurs between mothers and sons, perhaps because the kinship recognition developed by the long, intimate contact between the pair discourages copulation. Long-term studies of red deer reflect a similar pattern of father–daughter incest.

In species where controls have evolved preventing father–daughter matings, brother–sister copulations occur. Among acorn woodpeckers *(Melanerpes formicivorous)* in the Hastings Natural History Reservation near Monterey, California, which have been studied over several generations, offspring are reared in communal nests. A family ranging from two to fifteen individuals tends a single nest, helping to incubate the eggs and later feed the hatchlings. Observers note that females never breed in their birth group as long as their fathers are alive. Likewise, males do not reproduce within the group until their mothers die or are displaced from their nest by an unrelated individual. Females usually migrate but some elect to stay at home and remain virgins, in apparent anticipation of their parents' demise in order to take over the nest. The incest prohibition seems to work on the offspring of the opposite sex. However, when both parents of sibling woodpeckers happened to die at the same time, the siblings mated without inhibition. Since double orphans are relatively rare (observers noticed the phenomenon several times, but by no means often), it may be that incest avoidance in this species is neither innate—that is, it is not some kind of inborn knowledge—nor imprinted soon after hatching but results from social regulation exercised by the living parents, regulations that end only with the death of the parent of the opposite sex. These brother–sister copulations are apparently infrequent enough so that no genetic damage has resulted and avoidance behavior has not evolved.

When inbreeding reenforces the occurrence of genetically beneficial traits, there is no problem. But among animals as varied as the elephant shrew, the giraffe, and the reindeer, the high degree of inbreeding significantly increases mortality rates. Cooperative research on cheetahs suggests very strongly that all the cheetahs newly removed from the wild in South Africa (or those having spent only one or two generations in captivity) are unusually close genetically. Analysis revealed 10 to 100 times less genetic variation than among other mammalian species. In addition, a study of cheetah spermatozoa revealed 10 times less concentration than in the spermatozoa of domestic cats; and of those spermatazoa, 71 percent were morphologically abnormal, as compared with 29 percent in cats.

The relatedness of the South African cheetah population was tested with an experiment in skin grafts on captive animals. Skin was removed from fourteen individual cheetahs and grafted onto both closely related and seemingly unrelated individuals. All fourteen cheetahs accepted the grafts. Rejection, we know, occurs when tissue types of the donor and recipient do not match; and such matches are much better between closely related individuals. The ease with which cheetahs accepted the skin of a strange animal reinforces the evidence that even unrelated wild cheetahs are extremely close to one another genetically, as close as siblings.

Coincident with this evidence of inbreeding is a very high level of infant mortality in both captive and wild cheetahs. When the Wildlife Safari Park in Winston, Oregon, where some of these experiments were performed, was hit by a case of feline infectious peritonitis in May 1982, the previously thriving cheetah breeding program was decimated as the viral infection spread. The genetic uniformity among the individuals left them all prey to the same viral agent. Thus, the researchers suggest that the apparent consequences of high genetic uniformity include great difficulty in captive breeding, a high degree of infant mortality in the wild as well as in captivity, and a high degree of spermatozoal abnormalities, all of which argues for the advantages of abundant genetic variation for successful adaptation.

Incest avoidance is the rule in most mammalian species, and it seems to be accomplished by a combination of early negative imprinting, as well as a pattern of migration by either adolescent males or females so that they are far away from their birth groups when they are ready to mate for the first time. Females in most species that have the chance choose a distantly related or an unrelated male to father their offspring.

The most frequently observed mechanism for incest avoidance in many species, from Belding's ground squirrels to mountain gorillas, is male migration. Usually the females of the species remain together, mother and daughters as well as juvenile sons, until the adolescent male is either pushed out or leaves his home territory voluntarily. As the male matures, the mother's response to him alters suddenly. Small orangutans that have traveled on their mother's back, nursing at will and sleeping alongside her in her night nest, are shunned as if they had never existed. They beg for their mother's attention for days, trying to regain her interest. The male finally accepts exile, leaving his mother's side and eventually the territory. Among lions, the same phenomenom occurs as the cubs grow larger. Young females are allowed to join the larger pride, while the males are forced to the fringes. They trail the pride for a while, then eventually wander off alone to find a new group.

In a few species such as acorn woodpeckers, chimpanzees, and African wild dogs *(Lycaon pictus),* the females are the ones to leave the natal group. They scout around for another group in which, as adults, they will not be prevented from producing young. While they are in the company of their fathers and brothers, they are apparently inhibited from making sexual overtures; but whether the inhibition comes from their male relatives or from their mother is unknown. Among wild dogs in the Serengeti, young females entering maturity emigrate if their natal group already has a breeding female, who is probably their mother; in this case it seems the pressure to leave is triggered by the breeding female, who is suppressing reproductive competition, as well as by the odor or sight of a group of sexually attractive males. The migrating female looks for a group that has no breeding adult female. Often whole packs of sisters leave together. But if they remain together in their new group, only one among them will breed. The other sisters will eventually emigrate again until each female has a group in which she alone will become a mother. Apparently mortality is high among these emigrating female dogs, which accounts for the high ratio of males to females in this species.

The incest-avoidance mechanism is not very strong among some species of primates. Father–daughter copulations have been observed frequently among chimpanzees and gorillas. Within groups of mountain gorillas, whose dwindling numbers have made inbreeding frequent, webbed digits (syndactyly) and wall-eyes (strabismus) have appeared repeatedly within one particular family group.

Whereas female or male emigration among these species gives the appearance of having evolved as an incest-avoidance strategy, fresh evidence suggests that the emigrating individuals may be investigating new territories in search of food and incest avoidance may be merely an auxiliary benefit.

Forced Insemination

The usual mating pattern in the animal world is one in which the female carrying an ovum that is ready to be fertilized courts and then accepts the intromission of a male bearing active sperm. Within some species, the two sexes scarcely seem to notice each other outside the mating season. Among many monogamous animals, the pair live as partners but without apparent sexual involvement. Wild beavers spend more time building and repairing their den than they do copulating. Yet here, too, when the time is right, both partners cooperate to mate successfully. The important word is "cooperate." Most females in the animal world cooperate with males to perpetuate their species.

Yet there are exceptions. The majority of female animals choose their mates and become parents. Among males, however, the situation is different. All male animals do not get to be fathers, especially among polygynous species, where a few mature males monopolize all available females. It is not surprising, then, that violence occurs in nature—violence between males competing for sexual access to females, and sometimes violence between male and female when the male's advances are unwelcome. The imposition by a male of his reproductive ambitions on an unwilling female occurs among some species of insects, birds, and primates.

Ornithologists have observed assaults on female mallard ducks *(Anas platyrhynchos),* ringdoves *(Streptopelia risoria),* pintails *(Anas ocuta),* and blue-winged teals *(Anas discors).* Among mallards, the assaults often occurred before either females or males were in season. These were what the biologist David Barash calls "winter rapes." Their purpose, according to the ethologist Jack Hailman, could not be for reproduction because the drake continues with a precopulatory display: nod-swimming around his already-paired victim. Hailman suggests that the male's aim is to entice the female away from her partner. Whether the purpose is seduction, or pure aggression, the results for the female mallards can be disastrous. They take every possible measure—biting, swimming away, and flying—to avoid the attack. In so doing, many females die. It is the same among the blue-winged teal in Manitoba. Paired females are pursued by single males, with turns of the head and repeated calls. But their "no" squawks are not acceptable responses. Often a mate succeeds in fighting off the aggressor. But when observers watched eight males descend on a mated pair, the female's efforts to escape were frustrated at every turn. In this instance the harassed female dived beneath the waters to escape, and was never seen to resurface.

The ethologist Hans Kruuk recorded a curious instance of assault among the spotted hyenas *(Crocuta crocuta)* he was studying in East Africa. In this species, in which females dominate most behavior, he watched a handful of males try unsuccessfully to mate with an adult female, only to be rebuffed time after time as she returned to her den to suckle her ten-month-old cub. In apparent frustration, one of the males then turned to the cub. Eight times in succession the rejected male mounted the screaming infant without inserting and ejaculated. Each time he assaulted the cub, he faced the mother hyena who had refused him and then went over to her.

Adolescent male orangutans have also been observed attacking, and penetrating, uncooperative young females. The assaulted females appear to dislike the experience and do not indulge in any of the foreplay that characterizes most orangutan matings. On the contrary, they make quick work of it, screech their dismay, and scurry off as soon as possible. As far as is known, none of these unwilling copulations has resulted in pregnancy.

Among scorpion flies, too, where insemination usually occurs after a courtship in which the male has offered food to the selective female, there are instances of sexual assault. In this case the male copulates without offering food. As with the orangutans, these forced unions do not result in insemination as often as voluntary ones do. Only half of the scorpion flies who were unfed before mating reproduced, although tests of the sperm count of the assaulting males showed them to be as high as those among the males whom the females accepted willingly.

Forced insemination in the animal world represents a breakdown of reproductive cooperation and compromise between the sexes—one in which females become the unwilling victims of more powerful males. But usually males and females accommodate one another's reproductive needs, despite their often conflicting self-interests. In the short run, the arrangement they work out, like that of Peter the Pumpkin Eater and his wife, may limit the sexual behavior of one or the other sex—as when females must accept vaginal plugs along with sperm from promiscuous males, or when juvenile males are exiled from their natal group. But in the long run, many such limitations on sexual behavior among animals have worked to the mutual advantage of the sexes and have ultimately contributed to the species' survival.

part iii

Motherhood

nine

The Stuff of Fables

ENOUGH FEMALES reproduce to fill almost every ecological niche of this planet with offspring, some helpless and dependent, others immediately ready to be off on their own. The amount of care mothers give their young varies from what appears to be self-sacrifice on the mother's part to selfishness that could outrage a sentimental observer. But sentimentality obscures our understanding of success in the world of nonhuman animals. On the whole, animal mothers do care for their young, each in her own way; it is, after all, in her genetic interest to do so.

Against what sometimes seems enormous odds, females usually manage to produce enough eggs that get fertilized and develop into healthy animals to perpetuate their species. This has become increasingly difficult in the face of the spread of human civilization. Species are disappearing daily, and an enormous number of animals are sorely threatened. In the past, cataclysms like volcanic eruptions and earthquakes rapidly erased some forms of life. Today's cataclysm is the blindness with which our own species destroys rain forests and other habitats at breakneck speed.

Yet even without the newest man-made obstacles—chemical pollution and destruction of the environment—the developing eggs or larvae of insects and fish as well as the helpless young of birds and mammals are sought after as food by other creatures in the water, air, forest, and savannah. All life forms are threatened with disaster as soon as they emerge into the world. Those individuals that survive as far as birth and then sexual maturity bear testimony to enormous powers of endurance—and good luck.

Some young make it on their own, like codfish, whose floating eggs are abandoned to the waves of chance. Others, like bears, grow protected inside a mammal mother, and even after birth are coddled through a long dependency. The codfish has only the tides to thank. But to achieve this single living offspring, the female has spewed out literally millions of eggs. She has expended—some say wasted, in the sense of evolutionary economy—great quantities of energy in order to provide perhaps a single carrier of part of her genetic code to the next generation. Another strategy employed by different species is more parsimonious. For instance, orangutan mothers produce one offspring about every seven years, which they nurse and nurture for four years until it is ready to live on its own. Between the codfish and the orangutan lie most other animal mothers, each investing a different amount of care at different stages of her progeny's development.

It seems obvious—lots of eggs with little care, as opposed to a single egg with lots of care—yet this understanding is recent. In the last few decades, studies such as those of John Maynard Smith have generalized the different reproductive strategies by which most animal species maintain relatively stable populations. Some, like lemmings, produce large numbers of offspring in some years and few in others. Other species seem to produce more males than females under certain climatic conditions. Biologists do not attribute to animals any conscious choice in the reproductive strategies they pursue, whether apparently selfish or selfless. Rather, it is assumed that those reproductive characteristics that are best suited to the environment in which they must operate will prevail.

Egg Care

Midsummer through midfall, at Heron Island on Australia's Barrier Reef, hundreds of tiny green turtles scarcely an inch and a half long emerge in the tropical moonlight from nests deep in the white sands, to crawl across what must, in their view, seem an infinite stretch of beach. Those that complete the journey plunge into the Pacific, completely independent and off to seek what life awaits them.

These animals are born able to fend for themselves. However, a great deal of maternal effort has already been made in their behalf during that

Green turtles cross an expanse of beach, heading for ocean waters immediately after hatching.

crucial part of life that began with cell division and ended when they broke through their leathery shells. The needs of one form of embryonic life differ from those of another. Many fish, snakes, and birds that develop in eggs outside their mothers' bodies live more precariously at this stage than the protected fetuses safe inside marsupials and mammals. Those in eggs outside their mothers have a better chance if their mothers have made an effort to hide or camouflage their shells. Some mothers go even further, tending these eggs after hiding them to ensure that a greater proportion survive.

Embryos inside all eggs must be able to exchange gases, or breathe. They take in oxygen and emit carbon dioxide. Fish eggs, of course, differ from reptile eggs, and they again differ from birds' eggs. The shells seem to have become tougher and more rigid as the vertebrate species they protect moved from water onto land. Yet all eggs are tempting food for some predator. The trick, as some fish have discovered, is to keep the eggs near an oxygen supply, and at the same time keep them hidden, and often guarded.

Some females go to great lengths to protect their eggs. Octopuses lay theirs in a cluster attached to a stone or reef where, once anchored, they can protect them until they hatch. In the tropics, freshwater cichlids swim on continuous guard against all predators. On land, some spiders stay near their eggs until they hatch, scaring off invaders, while other egg-laying animals pick up their fertilized eggs and somehow carry them along as they wait for the young to emerge. Care and treatment of eggs makes a huge difference in the proportion of live offspring. Some females put all of their effort into hiding the eggs, while others amortize their efforts over the whole period of gestation. Some do both, burying their eggs with great care and then remaining on guard—just in case.

Those that put all of their energy into building the perfect nest include the green turtles of Heron Island. When they have reached five and a half feet in length and are mature, they make their way back to the coral cays on which they hatched. Coming in on a night tide very like the tide that took them out to sea, the egg-laden females lurch onto the sand, where they excavate large holes. Digging up with their front feet and patting down with the rear, they deposit their eggs inside. Observers estimate that every gravid turtle shifts at least a ton of sand each trip. And she returns to lay more eggs at intervals throughout the summer. Depleted, she will remain at sea again for three years. Meanwhile the turtle eggs are on their own. The temperature of the sand and the regular tides incubate them for approximately ten weeks. Then under another full moon, hundreds of tiny turtles break through their shells and, emerging with their heads toward the surface of the sand, they continue to push upward until they see the moon and begin the life cycle again. These cold-blooded animals rely on the outside world to regulate their body temperatures. Likewise, their eggs do not usually require any more heat than they can obtain from the environment around them. Nor do they need any more oxygen than is available in the sand and can permeate the leathery shell.

But some eggs, like those of the angelfish *(Pterophyllum scalare),* need a renewed supply of oxygen. Half a foot long, this silver-and-black-striped Amazonian cichlid fastidiously cleans the area around her eggs by removing bits of debris with her mouth so that fresh water can circulate, oxygenating them. Other females, such as the jewelfish, aerate their eggs by fanning them with their fins throughout the entire incubation period.

Among egg-guarding fish, there is a small gray genus in Mozambique, the *Tilapia,* that are *mouth-brooders.* The female picks up both her own freshly spawned eggs and her mate's sperm with her mouth and then retains the fertilized eggs between her closed teeth. She eschews food throughout the entire five weeks of incubation in order to avoid swallowing her developing young. After the fry hatch, she still keeps watch, letting them swim about at will until danger threatens. Then she somehow warns them—observers have not been able to identify the signal—but the fry respond as if she had pulled a retractable string that draws all of them back into the haven of her mouth.

Mouth-brooding fish live in a startling variety of species in the high African lakes and have become a favorite subject of animal behaviorists. In some species the male "mouths" the eggs, fasts, and protects the fry; in others both parents share the job, while in still others the female guards them by herself. Curiously, when the mother is the sole guardian, the fry family remains together for much longer than when the father plays or shares this role.

Where fertilization is external, as among most fish and amphibians, egg care by either parent is, in fact, less the rule than the exception. But it is not rare. Among toads, more often it is the mother who inconveniences herself physiologically to look after her unhatched eggs. Although toads do not carry the developing fetuses within themselves, feeding them directly as mammals do, many female amphibians shuttle their unborn young about in ways that are reminiscent of the complex internal feeding and protection system unique to mammals. There is, for instance, the marsupial frog, a tree dweller that lives in Mexico and Guatemala. Her red eyes, green sides, and creamy underbelly make her hard to spot amid the dappled greenery of the forest. The male slips the eggs that he has just fertilized into two hidden pouches on her back. He leaves her looking "pregnant" in reverse as she hops about with her developing tadpoles atop her. Safely enveloped by the liquid on their mother's back, they become tadpoles, then lose their tails and become frogs, while still inside her pouches. When they are ready to hatch, their mother reaches up with her hind foot

A female marsupial frog with eggs about to hatch on her back.

and pulls open the slit-like openings so that the young, fully developed froglets can hop away.

The Surinam toad *(Pipa pipa)* also carries her eggs dorsally, but without an organ that has apparently evolved especially for the purpose. Instead, as her mate sticks about sixty eggs onto her back, her skin slowly swells to cover the eggs, encasing each one in a separate pocket. They, too, pass through their tadpole stage under cover until, when they are toads, the skin of each little pocket opens like the lid of a jar, and they squiggle free. Although the small toads are attached to their mother by her own skin, there is no apparent contact between the mother's body and the growing offspring. She does not nourish them at all. That has been taken care of by the generous yolk in her eggs. Yet having them safely on her back enhances the probability that all of her offspring will survive as far as hatching.

Sometimes it is a question of providing the right temperature for the embryos rather than a mobile home. The fetuses of warm-blooded animals—birds and mammals—must be incubated at a steady temperature before they are born. But incubating eggs and larvae is not exclusive to them. Many insects, both social and solitary, take care of their embryonic young. The necrophorus beetle, so named for the rotting, putrifying flesh it eats, chews the decomposing mass into a ball that she then covers with her eggs, using her own feces as an adhesive. Even after the larvae hatch, the mother beetle remains atop them on the globe she has provided. Soon she feeds them directly from her own mouth with a brownish liquid she has manufactured from the rotting flesh. It is not known whether the female actually incubates her young or simply protects them with her body for as long as they depend on her for food. In either case, beyond this early relationship there is a total lack of social organization among these beetles. After its birth, the beetle will live solitarily for the rest of its life. In other words, the mother–offspring connection, at least among these invertebrates, is neither social nor communicative but is rather a physiological response in which the young are not treated, as they are in most vertebrates, as biological extensions of the mother. However, although the eggs might well develop without the bonus of a warm, protective body, the mother's effort assures the offspring a better chance at the start.

The female python is as cold-blooded as any frog or salamander, but when she is in charge of a litter of eggs, she coils her body around them, and at the same time her own body temperature rises. She is, in fact, incubating them, although python eggs do not need this additional heat to develop into baby snakes; in laboratories, python eggs will hatch in colder temperatures. But the heat hastens their development in the wild and probably allows the small snakes to enter the world at a time when food is plentiful and natural predators scarce. It is likely that the rise in maternal temperature evolved after the mother python began coiling her body around her fertilized eggs simply to protect them in a manner reminiscent of other reptiles. Female long-tailed salamanders in the eastern United States, for example, grow to only seven inches in length, of which almost

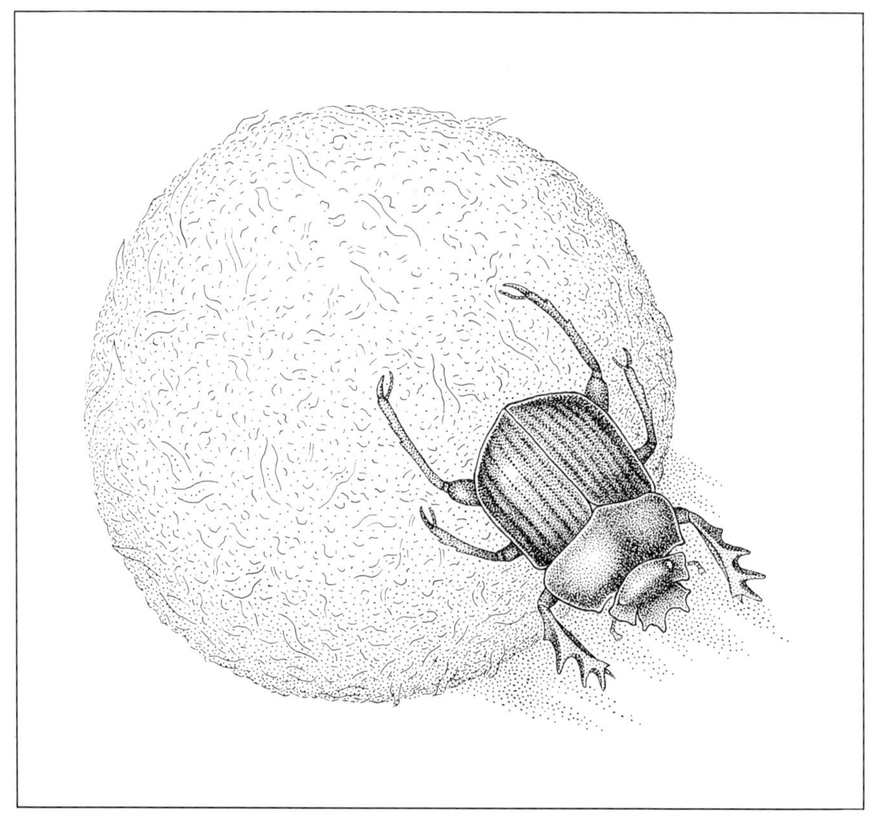

The necrophorus beetle constructs a dung ball into which she will lay her eggs.

four inches is taken up by tail, so that they resemble snakes in shape as well as in the way they coil around their incubating eggs.

Another protective group of reptile mothers are African crocodiles and New World alligators. Although both are cold-blooded egg layers, their nurturing habits do not fit conveniently into the evolutionary picture. Neither alligators nor crocodiles brood their eggs, at least not in the way that pythons and most birds do. But crocodiles are nurturing. Females grow to eight or nine feet in length and wait six to eight years before they lay their first clutch. When this finally happens, the new mother digs a nest in a sunny spot, drops her eggs inside, and covers them with dirt. Then she retires to the nearby shade to rest, but not to relax her guard. She maintains watch for three months, never leaving her post, not even to eat. Deep under the sun-dried dirt, the growing crocodiles inside their shells begin making loud noises and with a special tooth begin breaking through the shells. Hearing the muffled sounds, the watching female crawls out of the shade and hurriedly digs through the clay to release her hatching offspring. She helps those still in their shells and then scoops the infants, each only six inches long, into her mouth. She transports them between her opened teeth to a new hole which she has just excavated in the shade, and over which she will keep watch for another six months.

Likewise alligators, New World relatives of Africa's crocodiles, have been observed by biologists in the Florida Everglades protecting their own small offspring. These two species are by no means similar to each other in most aspects of child care, but both seem to combine the choice of a particular nest site with maternal vigilance to ensure the continuation of their species.

Birds, as contrasted with reptiles, are warm-blooded and their eggs need incubating. Most female birds devote a large portion of their adult lives to sitting on eggs, turning them from time to time to see that they are equally warmed and in every way provided with the right thermal environment. Unlike the developing pythons and beetles, warmth is neither a frill nor an accelerator to the young avian but a necessity without which no chicks would hatch. Birds, by definition, are covered with feathers. An outgrowth of skin, feathers are similar to mammal hair and nails in that they have no nerve endings, and once grown, though still attached to the bird's body, become dead tissue. Feathers help those birds that fly to stay aloft, and as a skin cover, feathers effectively insulate the bird from extremes of temperature. Yet the same feathers that protect the foraging bird from the cold may handicap her in transferring heat when she settles in to brood her eggs. To compensate for this feathery armor, evolution has provided many species with a brooding patch, a space low on the belly that becomes temporarily bald when the bird is brooding. This spot, which is closest to the eggs when the female sits on them in her nest, allows heat transference from the parent's body to the eggs. A peek at any dove or pigeon in urban North America steadily sitting on its eggs for about twenty-one days in any city park shows this remarkable brood patch.

Since brood patches would be a thermal luxury in the ice plains of Antarctica, King penguins have a way to heat their eggs without sacrificing any of their thick, waterproof feathers. These birds have a brood envelope, a fold of skin on their feet in which they store the egg as it develops. The male penguin remains still, hugging the egg, and like the female crocodile in the tropics, does not eat for the entire fifty days of

A female python incubating her eggs.

incubation. Fortunately, or by necessity, King penguins are monogamous and have loyal mates. As soon as the penguin chick hatches, the mother takes charge and allows the by-now starving father to set out for what is often a very long trip across the channels to feed himself and bring back food for the offspring, after which the parents take turns baby-sitting and foraging.

An even closer approximation to the relationship that mammal mothers have with their offspring occurs among some invertebrates—those in which the mothers not only carry the developing embryo inside their bodies, as true mammals do, but also manage therein to feed them. The embryos of velvet worms that live under rotting logs grow inside the mother worm. Here they develop in a way that Wolfgang Wickler points out is by no means primitive. They are fed through a membrane very much like a mammal's placenta, and are born well-developed, virtual copies of their parents.

Alligator hatchlings in the safety of a nest provided by their mother.

A female king penguin in charge of her month-old chick.

Members of the family Goodeidae live in the fresh inland waters of central Mexico's highland mesas. These small fish also nourish their unborn through an elaborate placenta-like structure. The embryos use up their small supply of yolk early in their development and then send out ribbon-like growths which transport food to them from their mother until they are born. And surfperches (Embiotocidae), which live off the California coast, have placental connections through the fins of the developing embryos and are truly viviparous. Closer still to the mammalian way of internal nourishment of its unborn is the North American garter snake. When pregnant, the female carries from three to ninety-two embryos inside her abdominal cavity and feeds them through a rudimentary placenta.

It seems as though some kind of internal nourishment improves the chances of embryo survival. The process appears convergently in different species with no lineal connection to each other. In evolutionary time, placental mammals appeared most recently. Among these animals, the placenta entirely replaces the egg yolk and shell. It is a specialized membrane that develops inside the uterus of the pregnant female and serves as filter, feeder, and protective envelope for the embryo. The placenta enables a mammalian mother to begin nourishing her offspring before it is born and separate from her own body.

Feeding the Young

Mothers that care for their young long after they are born continue to do some of the same things they did for them when they were still incubating or pregnant. They feed and defend them from both the elements and predators. Many of them are *altricial,* a term describing a variety of dependencies, usually meaning that the newborn infants cannot see, defend, or feed themselves. Altricial young need help, and they often get it from their father and other individuals as well as their mother. Altricial young have a long way to go between birth and independence.

Yet many mammal and bird babies are born *precocial,* a term meaning that they can pretty well take care of themselves. Most ungulates have legs that are sturdy enough to support them within moments of birth. They meet the world with open eyes, their bodies insulated by fur. Precocial birds are born with feathers, and seem like miniature adults. Yet few of them hatch and then depart immediately, despite full plumage and sharp eyes. Mammal infants can never simply peer bright-eyed into the world and scoot away. They depend on their mother's milk no matter how precocious they may seem.

Mammal mothers feed their newborn with milk they manufacture in their own bodies, in the very mammary glands that give the class its name. This method of feeding is employed by all mammals, including those special mammalian families, marsupials and monotremes, that are peculiar to Australia. Mammal infants differ among themselves as to the quality of milk they drink, from the fat-rich fare meted out to young spermwhales to the thin, watery stuff enjoyed by infant kangaroos. But whatever the proportions of water, sugar, and fats, mother's milk is a complete food for the newborn. It contains a full supply of proteins as well as special antibodies that protect the young mammal from a variety of infections. Each mammal's milk is unique, although it is possible, as Western society has learned, to substitute cow's milk for that of other animals, such as humans, cats, and dogs, that for some reason have been deprived of their own mother's milk.

The mammary gland evolved from a gland initially used to produce sweat. Evolutionists speculate that it became active in pregnant females as a way to keep their offspring moist. Part-way along the evolutionary trail is the platypus, an egg-laying, warm-blooded monotreme that inhabits Australian streams. The platypus's milk ducts are not confined within fatty tissue like a breast but instead lead directly to the surface of the skin, where the liquid simply oozes out with no nipple at all.

Nursing mammals differ from each other in the ways they drink. The infant platypus, unlike other mammals, hatches from a leathery egg and then licks the milky surface of its mother's abdomen. The small kangaroo latches on to one of its mother's nipples with such vigor that the nipple reaches deep into its mouth and enlarges, forming a little button that locks the tiny animal in place. That nipple becomes the infant's property. As months pass and the small kangaroo grows fur and becomes strong enough to hop in and out of the mother's pouch at will, the milk it drinks gradually becomes fattier. But even after a year has passed and the youngster is almost independent, when it returns for food it reaches for the same nipple that seemed to attach it to the mother so many months before.

A kangaroo "joey" nurses at a teat inside its mother's pouch.

The giant infant spermwhale, almost fourteen feet long at birth and weighing 1,700 pounds, has a different modus operandi. The mother rolls on her side just below the surface so that the infant can breathe through its blow-hole while it feeds. Whale babies do not have mobile lips so they cannot suck. Instead they latch onto the mother whale's nipples, which she removes from a streamlined pouch on her underbelly and through which she pumps the milk into their open mouths.

Mother orangutans and gorillas are helpful, too. Their offspring need to be directed to the breast and even placed on the nipple. They are very different from young piglets or puppies that race to find a place on line and suck heartily alongside half a dozen brothers and sisters.

The frequency of feeding and the length of time the young continue to be dependent on their mothers vary. At one extreme is the pilot whale, which may continue to suckle for seventeen years; at the other is the tiny tree shrew of Indonesia and Malaysia, which feeds one or two tiny offspring to the bursting point, then disappears for forty-eight hours before giving them another fill-up. Some mothers, like baboons, leave their breasts always available so that the infant can suckle on demand. Others, like porcupines, keep their sharp quills alert and only after persistent demands will sit back to expose their defenseless, benippled underbellies to their whimpering offspring.

Other species besides mammals, monotremes, and marsupials have evolved mechanisms to feed their young with substances produced in their own bodies. The necrophorus beetle with its brown nectar is one example. Others include, among birds, female pigeons that produce a special food for the first four days of their nestlings' lives. Although different from mammal milk, crop milk is also stimulated by prolactin, the same hormone that triggers the secretion of milk in mammals. Crop milk does not develop in a special gland but in a unique organ, the pigeon's crop, an expansion of its esophagus that operates for only the short period immediately after hatching. For these few days the crop functions as a factory in which female pigeons turn the grain they feed upon into a thin, milky substance.

Among fish, too, there is a species that produces an analogue of milk, in this case another member of the ciclid family, the South American discus *(Symphysodon discus).* Discus secrete mucus all over their skins. Monogamous, both father and mother look after the young and both secrete the

A lowland gorilla female examines her newborn infant.

nutritious substance. Although the fry only depend on it for the few days before they are able to eat exactly the same food as their parents, without it during that period they would die.

Some insects, unlike most other animals, continue to manufacture food throughout their whole lives, using it to feed both embryonic young and full-grown adults. Perhaps the most familiar insect food is honey, which bees produce to feed their larvae in special cells within the hive.

A great variety of animals process what their young can eat, breaking down portions of the foods that adults of their species eat into a soft mince. Among the koala, marsupials of Australia, the mother maintains a monotonous diet of eucalyptus leaves which she digests in her caecum, a pouch that branches off from her intestine. Then, after her offspring have stopped suckling, pureed eucalyptus leaves pass through her anus at special times of the day (usually between three and four o'clock in the afternoon) in a way that keeps the food totally removed from her excrement. The tiny youngster eats this mush until it is old enough to pick and chew its own eucalyptus leaves. The majority of termite species simply eat one another's excrement. Termites apparently eat very inefficiently, if what they pass out still contains enough nutritious material to satisfy another termite. Nor, apparently, are termites at risk from the harmful bacteria that make most mammal droppings unhealthy to others of their species.

Most fish do not feed their young, because the adults are so much larger than their tiny fry that they do not share the same diet. The adults often eat smaller fish, while the fry feed on microscopic plankton. Here, too, there is an exception with the mother fire mouth. After clearing a spawning site, both mother and father protect the eggs, and when hatching is imminent, both parents dig pits and move the fry to safety in them while the mother demonstrates to the youngsters what to eat by capturing smaller fish and pre-chewing them for the immature offspring. In like manner, some mother spiders regurgitate small food droplets to their young, and some cockroaches also feed their offspring with a predigested mixture of their own.

Females of many species of gulls fly out to sea and capture fish. On returning to the colony where they have left their offspring, they regurgitate the partially digested dinners to their chicks. Pelican mothers off the California coast are noted for their enormous gullets, from which their young extract whatever partially digested fish they can find. Providing

An infant pelican feeds from its mother's gullet.

food for an infant bird can be very tough. Researchers studying King penguins on South Georgia Island in the Falklands reckon that parent birds go to considerable depths to get the necessary squid. Diving down as far as 160 feet, the penguins must eat two pounds of squid to keep up their energy for every pound they bring back. Averaging 144 dives a day, penguins are gone for four to eight days at a time to forage for the food their chicks need to survive.

The sight of a mother bird gently dropping a worm into the open beak of her newborn chick is familiar to everyone who has seen a nature film. In this case, the mother finds the kind of food for the youngster that the youngster will eventually need to find for itself. Carnivorous birds, like owls, hawks, and eagles, bring back whole animals, freshly killed (or not quite dead) to their nests, where they tear off small pieces of bitesized flesh for their offspring to munch. Flesh-eating mammals have gustatory needs similar to these birds of prey and raise their offspring in remarkably similar ways. Hiding the young in a burrow is akin to stashing them out of the way in an aerie. The mother hyena, fox, or lion will capture and kill its meal, then either chew it on the spot and regurgitate some of it back at the den or, if she is close enough, will simply carry back bits of flesh for the hungry brood.

Outside the world of vertebrates, insects such as the scarab beetle leave their offspring alone to hatch inside balls of dung, which provide the larva with a rich meal as soon as they are able to feed. Some mother spiders leave wrapped-up parcels of live insects for their newly hatched to chew. The *Eresus niger* mother feeds her spiderlings at first by regurgitation. Later, when she runs out of food for lack of an opportunity to forage, she offers herself, so to speak; her sixty-odd offspring suck her until she dies, and even afterwards they suck her dry.

Even those mothers that do not actually deliver morsels into their youngsters' open mouths do help their newborn to eat. Mother chickens do not drop grain into their chicks' beaks, nor do their peacock cousins. But they nod and point their heads and indicate by gesture and example what the youngsters are to eat. Colobus monkeys in the African rain forests hop about with their mothers, as do small gorillas on Mount Virunga in Zaire; and while living on milk, they watch and remember which leaves and berries their mothers choose. As they develop teeth, the young imitate her and follow a proper diet. North American lynx cubs share a lair with both parents for several months, nursing from their mother until they are old enough to follow her onto the tundra to learn to hunt. In preparation, their mother brings back fresh meat and lets the youngsters suck at it, then nibble, until they have developed a taste for the kind of food they will have to kill for themselves when they are ready to go out hunting. By the end of their first year, all young lynx have to do this on their own.

Protecting the Young

Feeding is not the mother's only responsibility. In the wild, as opposed to the protected environment of zoos, mothers must keep eye and ear alert to a multiplicity of dangers. These females must protect their new investment even as they suckle or pick berries. For these mothers, a well-fed calf is a luxury, while a protected one is a necessity if they are going to continue being mothers at all.

Looking again into the watery world of fish, we see that while few species have evolved a method of direct feeding, a great many have developed a protective system. The tiny anemone, which establish their home inside a sea anemone, lay their eggs on a nearby rock, and both mother and father guard them by hitting with nipping bites any untoward fish that ventures too close.

Among birds that live in colonies and hatch offspring together, such as Arctic terns, the arrival of an enemy triggers the adult mothers to team up and jointly attack. Depending on their species, these colonial sea-birds might fly after the villain, or even surprise it by dive-bombing assaults that may maim or even kill the would-be diner. Mobbing a predator can serve a double purpose. It can put off the individual antagonist and at the same time cause the kind of noisy distraction that will let the offspring hide while the predator and any others in its wake concentrate on self-defense.

Many animals, of course, prefer flight to fight. Sensing danger, they signal their young. Wild zebras and deer take off when they are uneasy. A leopard mother raises her tail high like a flag and uses it to lead her young. Other species such as swans and chimpanzees hoist their babies onto their backs and carry them to safety. But if they have not moved fast enough, or soon enough, mother antelopes and elks that are usually docile animals will put up a surprisingly hard fight with hoof and horn to defend their calves.

When mothers watch over their offspring alone, they usually opt to hide or disguise them, rather than to wait until a predator arrives. The shy bushbuck eradicates all evidence of the newborn. Not only does she lick up the afterbirth, but she also removes all the baby's urine or feces. Among red deer on the northern Scottish island of Rhum, occasionally twins are born. The mother hides each in a separate place and disguises evidence of both from each other and any enemy.

If hiding an infant fails, a bird like the nighthawk feigns a broken wing to divert a predator's attention from the helpless infant. Other animal mothers distract predators with warning calls. Adult female Belding's ground squirrels squeal, focusing attention on themselves instead of the helpless young. There is some risk involved, for the mother exposes herself to danger which she cannot always evade. Yet it seems that in most instances the warnings protect the offspring without sacrificing the adult.

Predators are not the mother's only worry. There are often threats from within her species, even within her own family. Mother lemmings always leave their nests when they want to mate, keeping their offspring hidden in the tundra. Hawks and owls frequently keep their mates away from the young, although they do accept food donations; some females even fast themselves as if to avoid inadvertently eating their own children. Male spiders, too, if not prevented, eat the newly hatched babies. This analogous situation in which cannibalistic fathers pose a danger to their young might explain why, in both predatory birds and spiders, females so outsize their mates. They need the greater size to defend their young.

Predation and murder is only one threat. Mothers also protect their infants against extremes of climate, which the very young, lacking fur or feathers, cannot do on their own. Mother birds warm their naked chicks by sitting on them. Large mammals such as giraffes also have special needs relating to temperature. The giraffe was misunderstood for generations and considered a "delinquent" mother in popular scientific literature until recent research called for a reassessment. Giraffes are "hider-type" ungulates that leave their young during the day, traveling long distances away from them. It once seemed that they had deserted the youngsters. But Vaughn Langman managed the difficult task of attaching radio collars to the necks of several giraffes, and tracked individual mothers and calves. He discovered that mother giraffes leave their young in cool, grassy places where they visit them daily to let them suckle. After about a month, they bring the calves into groups, where other females act like "babysitters."

Female giraffes with a newborn calf.

While most of the females travel to browse for leaves, one or two remain behind to stand guard. The giraffe's unusually long neck presents a problem in thermal regulation that only time takes care of. The mother giraffe cannot let her young expose itself to the full sun before it is about a year old and mature enough to regulate its own temperature. Before that, the mother not only guards and feeds it but makes sure it stays in a cool enough spot so that it does not succumb to heat.

Teaching the Young

The line between innate and learned behavior is unclear throughout the animal kingdom. As we have seen, this is especially apparent when it comes to eating. Although some species, like the discus fish, suck instinctively, others have to be taught to take the nipple, and must be shown by example which particular foods to eat. Most animals that are not born with instinctive knowledge about how they live can learn only one way—by example. So while they are being fed by nurturing mothers, they are also learning what foods are good for them to eat, and how to get along in the social group they have more often than not joined. If they are female, they are also learning how to prepare for their own offspring should they become mothers, too.

While birds intuitively know how to fly, it is clear that in many species the mother gives demonstrations, teaching her fledglings the secrets of the air. Monogamous birds, those that will eventually pair off for a good part of their lives, need to learn the special songs of their species. While the general pattern of a song is apparently inherited by the young bird, the actual notes—the particular tune, so to speak—is learned by living with an adult of the species.

The degree of deliberate teaching that goes on between mother and child in the animal world is hard to pinpoint. Most learning is by example. However, the efforts of a mother cheetah with her cubs show that in this instance the feline mother is quite carefully going through the rules with the cubs over and over again. Cheetah cubs will be left on their own within a few months, and they must know how to hunt if they are to survive. The mother captures prey and brings it back to her den—alive and in a strangle hold. She lets the cub actually make the kill. The cubs may be weak and botch the task but she does not interfere.

Among social primates, baboon daughters learn the structure of their group from their mother. They learn to pay a kind of homage to a strong male leader, and to respect those other females above them in the social hierarchy. While playing games, small primates mimic adult activities and learn the right way to behave in their own groups. Although stress contributes to pathological behavior in zoos and primate centers, all captive animals are not victims of these pressures. Often they simply do not know how to behave because they were captured as infants and have had no adults of their species as models. Experiments with films, and in particular psychological conditioning, as illustrated by an experiment at the San

The training doll used to teach Dolly Gorilla how to be a caring mother.

Diego Zoo's Wild Animal Park with a captive-reared female gorilla, demonstrates that for some species mothering is a learned behavior. Dolly gorilla neglected her first-born in 1973; he had to be rescued and placed in an incubator to save his life. Eager to rear emotionally healthy representatives of this endangered species, zoo administrators resorted to the help of human psychologists. When Dolly became pregnant the next year, she was given a burlap pillow and conditioned to care for it as she would a baby gorilla, her instructor showing her how to cradle it gently and offer it her nipple. Seven weeks after the start of this conditioning, Dolly gave birth to her next infant. She nurtured it successfully, and did the same with succeeding offspring throughout the next decade.

ten

A Mother's Own Interests

THE WORD "mother" implies caretaker, yet there are enough examples in nature of mothers jettisoning their offspring to suggest that the word should apply only to the act of laying eggs or giving birth. Self-interest seems to motivate all behavior, female and male, and it is in the mother's interest to ensure a new generation by any method that works. According to those who define an animal as a gene's way of perpetuating itself, the individual parent's only real concern is the continuation of its genome, the total genetic package of the animal. This concern does not necessarily imply unlimited devotion to each individual egg, embryo, son, or daughter produced.

There is a school of biologists that discusses parental strategies in an economic idiom. They write about an animal mother's "investment" in the first offspring, as opposed to what she will have left to invest in those that follow. Animal parents make "unconscious choices," the school maintains, "deciding" at some point when to "cut their losses" and try to recoup next time around. This kind of discussion anthropomorphizes animal behavior in the modern economic idiom more subtly, perhaps, than Victorian sentimentalization, but it nonetheless projects human motivations onto nonhuman animals. Yet these approaches to animal behavior, while somewhat simplified by virtue of their mathematization, do highlight the evidence from observation that the interests of mothers and their offspring are often at odds. Robert Trivers pointed out in 1974 that a parent and offspring have conflicting interests in terms of reproductive ambitions. If allowed, all offspring would demand total and permanent maternal devotion. So a mother must weigh the advantages of supporting individual offspring as opposed to helping others, older siblings perhaps, or those offspring which she is still able to produce, perhaps under conditions more favorable to survival.

Bonding versus Abandoning

In the key moments after birth in many species, a special relationship between mother and child is established. The mother learns to identify her particular offspring and wants to help it at the expense of all other infants of her species. At almost the same time, the infant learns to recognize its own mother, to seek out that mother and not be satisfied with a substitute. This mutual reaching out for particular individuals is crucial to survival in these species. In terms of evolutionary tactics, it is clear that this attachment, or bonding—perhaps an evolutionary antecedent of love in *Homo sapiens*—is necessary for the preservation and nourishment of individuals within a group.

Observers are fascinated by how and when attachments happen. Mother harpseals can be seen nosing their pups almost immediately after birth and soon identify their own, apparently by smell. Orphans in this species never survive because, try as they might to reach an accessible nipple, the mother always knows the pup is not her own and chases it away. Northern fur seals in Canada listen to their pup's squeak and, after returning from feeding at sea, will wallow among a sprawl of over a hundred noisy pups and pick out their own by ear.

The psychologist Susan Kingsley studied mother–infant pairs of orangutans at the Regent's Park Zoo in London and saw that the mother kept eye contact with her baby, creating a visual bond between them that held firm. Mothers distinguish their own young very soon in most mammals. But the young's recognition of the mother, it seems, comes somewhat later. Once the infant recognizes its mother and demands her attention, the maternal bond that was already forming is reinforced.

Experiments with sheep reveal that stimulation of the vagina during labor, or artificial stimulation with a rubber bladder inserted into the uterus two hours postpartum, triggers the production of oxytocins. These hormones apparently encourage the new mother to devote herself to caring for her own lamb or, in the case of artificial stimulation, to adopt another lamb toward which she had previously been aggressive. Suppressing the mother's sense of smell while she is giving birth prevents her from identifying her own offspring; consequently, since she has the need to suckle, she will willingly suckle a stranger's offspring. Experiments with mice lead to similar conclusions. There seem to be connections in some species between vaginal stimulation, the mechanisms of smell, and bonding that are only beginning to be understood.

However attachments begin, they endure for months or years, depending on the species. Yet in the majority of species, the attachment ends almost as abruptly as it begins. Some mammals wean their young in a manner that is apparently painful to the youngster. The mother rejects the youngster as the calf or cub struggles to return to the nipple.

Relations between mothers and their male offspring are often different from those between mothers and daughters. In species where females remain together but young males emigrate, the bond between mother and son may be extremely close for a short time, then abruptly end when the male approaches maturity and exile. Lionesses evict their older male cubs from the den, but not from the pride, within three and a half months of their birth. Likewise, adolescent male orangutans—between eight and ten years of age—are forcibly chased from their mother's sleeping quarters and her side without much warning. Just as attachment seems to begin with the mother's highly sensitive responses to her own offspring, it is usually ended by the mother when she is either expecting a new family or merely senses the sexual maturity of her offspring.

In many species, however, the attachment between individuals endures. Observers of elephant herds have been awed by the apparent affection and attention to personalities that color the social lives of the females that live together with as many as three generations in one group. Mammals are not alone in their apparent personal preferences. G. P. Baerends and his wife, J. M. Baerends Van Roon, observed the behavior of cichlid fish in Lake Tanganyika in 1950. They noted that while the young leave the family group as they grow in size and form nonreproductive bachelor groups, they maintain social contacts. Living apart from their parents and their siblings, the young fish are visited frequently by family members. These relatives are neither foraging for food nor seeking to mate with them, but are just swimming back and forth frequently as if for the sake of visiting.

Stephen Emlen, an ornithologist, has chronicled the complicated social relations among the white-fronted bee-eaters in Africa. Several generations live together and help raise each other's offspring. Most remarkable, however, is the amount of visiting that occurs between individual birds, not for any apparent reason except to get back together.

The bonding that develops between parents and offspring can end abruptly with the death of the mother or infant. Jane Goodall has chronicled in great detail the lives of the chimpanzees at the Gombe Stream Reserve in Tanzania. Among all the relationships she describes, perhaps none is as poignant to us as that of the matriarch Flo and her many offspring. When Flo died, her adolescent son Flint revisited the nests they had shared and the places they had rested. Then gradually he wasted away and died.

The tie that binds a mother to her offspring is established at different

An infant elephant seeks protection amid a forest of adult legs and trunks.

times in different species. For instance, there is no bond between a marsupial mother and young at birth or even immediately afterward. This enables the marsupial mother to neglect, without suffering from stress, an offspring that would probably not survive. Red kangaroos live in an uncertain environment to which they have adapted with remarkable flexibility. The red kangaroo fetus develops from a fertilized egg that the female may have retained dormant for as long as several months. The fertilized female apparently waits until the food supply or climate signals a promising time to allow the blastocyst—the ball of cells that will develop into a fetus—to leave its state of suspended animation and move into her uterus.

Further cell division proceeds, and after 31 days the tiny bean-sized fetus emerges. It is still in its birth sac as it passes through her vaginal tract in the same manner that more developed fetuses emerge in true mammals. Unlike a placental mother, however, the kangaroo does not assist the delivery. To the observer she gives the impression of being unaware that anything unusual is happening. While she gazes at the passing scene, her young make their way on their own power up her stomach and into her pouch. Only one offspring is usually born at a time, although twins are not rare. But there are four nipples waiting in the mother's pouch. If the small fetus should fall to the ground before it completes its journey, it is doomed. The mother never attempts a rescue. Twins are even smaller than single babies but large enough so that once inside the pouch they compete for space or food. Only one fetus usually survives the more than 200 days of pouch life required for independence.

In the American opossum, also a marsupial, the lack of apparent maternal bonding takes a different form. Opossums produce as many as eighteen embryos at a time, although there is only nipple-room for seven. The system clearly anticipates high embryo mortality: however, if only one manages to complete the journey to its mother's pouch, it cannot survive. The mother opossum needs the stimulation of at least two suckling infants to keep her milk flowing. Yet the common Virginia variety will not bother to retrieve a young embryo that has become detached from her nipple; she will allow it to die. When female opossums were encouraged in laboratory tests to retrieve their young, they were totally unresponsive. Evidence that this indifference resulted from a lack of bonding comes from other tests in which these same mothers willingly, or perhaps obliviously, accepted young from other litters into their pouches.

Hardy marsupial infants that do manage to reach their mother's pouch grasp a nipple in their mouths and hang on tightly. The nipple grows along with the infant so that no amount of jiggling can dislodge the infant from its milk supply. The connection between offspring and nipple is so tight that for a while some naturalists believed them to be temporarily fused. It would seem likely that by this time mother and infant would be bonded emotionally as well as physically. Yet this is not the case for the red kangaroo. Its infant—more than 200 days old, now grown large and called a "joey"—spends much of its time outside the pouch, accompanying its mother and the herd she is part of as they browse in the countryside. The joey still nurses from nipples inside its mother's pouch, often from a standing position on the ground beside her. At this stage, it is no longer drinking the same milk it did when attached to the "baby"nipple. Just as

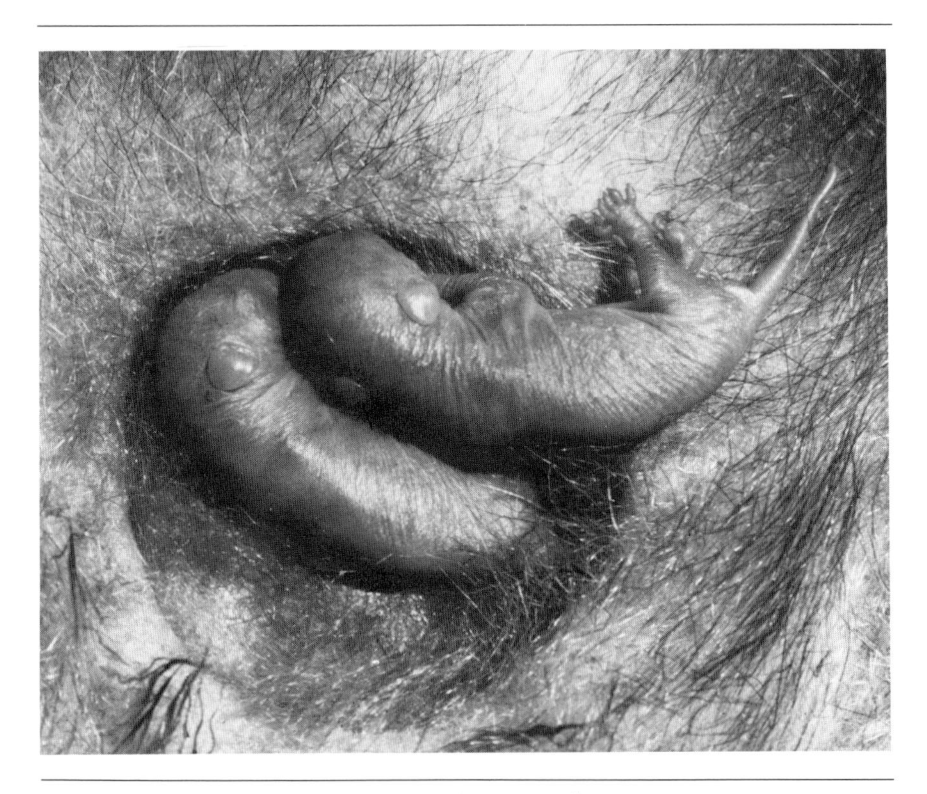

Infant opossums nursing in their mother's pouch.

the female kangaroo is able to control two separated stages of fertilized ova simultaneously, she is able to provide offspring at different stages of maturity with separate foods. There may be a younger infant inside her pouch full time by now, sucking thin milk while the joey drinks a richer formula.

When threatened, the joey climbs inside the pouch, often alongside its tiny sibling. But if the mother sees an ominous predator making too much headway, she will lighten her load by throwing out the heavier offspring. If this happens as she is hopping at the rate of thirty miles an hour, the jettisoned joey may roll a considerable distance. If it is old enough to have fur, and to eat solid food, it may recover and take off in a random direction. If it is still suckling exclusively, it will probably be caught by a scavenger before it starves to death. The female kangaroo has had a long period of time to bond with the joey, but her sense of self-preservation clearly dictates that she cut her losses, even after many months. She leaves the joey to its fate and escapes with the younger baby in her pouch and perhaps a blastocyst, which she can now allow to develop and replace the youngster she has just discarded.

The Australian outback is a precarious place to raise offspring on two counts: predators threaten all animals, and extended droughts often dry up most food sources so that only the hardy survive until it rains. So that the female kangaroo can thrive in the face of this uncertainty, evolution has apparently allowed her to have a steady supply of replacement offspring in lieu of the tight bonding that keeps most mother birds and mammals caring for their young.

Marsupials are not the only Australian mothers that abandon offspring when danger threatens. The flightless giant emus never seem to develop nurturing gestures toward their young, either. Laying their eggs in nests built by their mates, they leave the male of their species to brood them and only return after the eggs have hatched. Female emus then join large, herd-like flocks browsing with their young. But when danger threatens, these females will abandon a whole generation of their own offspring and set off at a run to save themselves. The parallel between the maternal behavior of Australia's birds and mammals is unavoidable: the unpredictable desert seems to produce species whose behavior is a response to hardship, making the bonds of motherhood appear to be a luxury of steadier climes.

Not only in Australia but elsewhere as well, there are females that show uneven talent as nurturers. In close proximity to hyenas, female wildebeests live in migratory populations where they give birth in regular calving grounds. Only minutes after birth, the calf is up on wobbly legs and looking for its mother's teats. Newborn wildebeests will follow anything that moves, and do not apparently imprint on their mothers until after they have had a chance to suckle. If they wander off before this, they are likely to end up as prey for the hyenas and lions that follow the herd. But once they have bonded, wildebeest mothers have been observed risking their life-sustaining connection with the herd by hanging back after crossing a stream to reunite with a lost offspring, whom they are able to recognize by the youngster's call.

Although lionesses have reputations as good mothers—and indeed they are, often attacking males in defense of their cubs even to the point of suffering wounds or occasionally death—their sacrificial efforts are concentrated on those sturdy and healthy cubs at least three months old who have been brought into the pride. At this point the youngsters join with agemates, probably cousins or even half-siblings, with whom they are fed and groomed in a communal creche. Yet only a few of the infant lions that are born make it to this place of relative comfort. A lioness goes off alone to give birth, and if after a few days only one or two cubs in the litter survive, she may abandon them. Ethologists have happened onto many orphaned lion cubs. They attribute this infant desertion to the lioness's instinctive sense of investment. Nursing a single cub will take as much of her time and energy as nursing half a dozen. By abandoning the few, she increases the chances of breeding again soon, at which time she may have a larger litter.

Occasionally a female appears to neglect her offspring when she is actually paying them minimal, though adequate, attention. Tree shrews of the Malaysian rain forests appear to be more involved in the monogamous relationship they have with their partner than in their relationship with offspring. The mother prepares a separate, hidden birth nest where she leaves her three or four infants almost as soon as they arrive to hurry back to the nuptial nest she shares with her mate. Being mammals, the babies need maternal nourishment, but the mother looks in on them only once every forty-eight hours to let them suckle. Alone, the tiny infants must trust to camouflage to elude predators, and to the slowness of their metabolisms to avoid hunger pains before their mother pays another visit.

A newborn wildebeest nursing and thus bonding with its mother.

Brood Parasitism

Females of several avian species have developed an ingenious variation on the phenomenon of maternal bonding. Some members of the cuckoo family slip their fertilized eggs into the nests of other birds whose nurturing instincts are, apparently, stronger. The behavior of these females has been labeled *brood parasitism.* The cuckoo mother becomes a maternal parasite of the foster mother, using the pied wagtail *(Motacilla alba)* or European redstart's *(Phoenicurus phoenicurus)* maternal energy to care for her cuckoo chicks.

The pressures that selected for the development of parasitism may have been multiple. The parasite mother may have been unable to find adequate food or cope with predators. These females removed themselves from direct maternal responsibilities, but not before ensuring the safety of their offspring. They lay their eggs in nests with "look-a-like" eggs so that the chosen substitute mother does not distinguish the changelings from her own. These females rely upon an instant bonding process, that intangible imprinting that links mother and offspring in most species and causes the two halves of this parent–offspring equation to depend upon each other. Bird mothers seem to recognize their own young, probably by smell or sight, very soon after hatching. The mother thus picks out those she believes are her own offspring right away, and the chicks look to her in response for nurture.

The Old World cuckoo *(Cuculus canorus),* a slender long-tailed member of the family Cuculidae, is familiar to city-dwellers and villagers across western Europe and into Asia. Evolution has left the female of this species with eggs that mimic those of her unknowing hostess. The wagtail unwittingly incubates a cuckoo's egg along with two or three she has already laid herself, blind to the "trojan horse" in her midst. For when the chicks hatch, the cuckoo baby inevitably kills its nestmates and then accepts all the food its foster mother manages to forage. There is, it seems, one instant during which the redstart will interfere with this process; it is the moment when the stranger's egg enters her nest. But the female cuckoo has adapted behavior to avoid this possible confrontation and does not sit on the redstart's nest to lay her egg. Instead she swoops in and swiftly hovers over it, taking only seconds to drop her eggs, which have an especially tough shell to withstand the impact. This is in contrast to the laying behavior of most birds, including the deceived redstart, that spend from several minutes to several hours completing the laying process.

One mystery, of course, is why the redstart continues to care for the baby cuckoo after the baby cuckoo has done away with her own young. Why have redstarts not evolved their own safeguards against the cuckoo's intrusion? It may simply be that there are many more redstarts than cuckoos and the mortality rate is not high enough to have prompted the adaptation of defensive strategies. The ease with which the redstarts accept the cuckoo, and consequently imperil their own offspring, makes it appear that the redstart mother is the negligent guardian, for her own chicks are forfeit. The parasitic cuckoo mother, on the other hand, is careful, having arranged that her offspring are well provided for by a good, if not discerning, stepmother.

Unlike the cuckoo, the African honeyguide *(Indicator indicator)* does not rely on her young to destroy the offspring in the nest where she deposits them. Instead, she lays her eggs with starlings, destroying the starlings' own eggs before she leaves. The honeyguide chick will hatch into the world with a special hook on its small beak. In the event that the starling lays more eggs after the mother honeyguide has departed, the chick will use its hook almost immediately to lacerate and kill any young starling that may appear.

Less lethal to its foster family is the African widowbird *(Diatropura procne)*. She exploits the finches in the family Estrildidae but does not go so far as to destroy the young. Rather, she is part of a complex deception to ensure her chicks' protection. Widowbirds have learned to mimic the finches' song, and evolution has provided the imposter chick with the same pattern of dark spots inside its gaping mouth as those on the newly hatched finch. The female thus regurgitates her meal of seeds into the changeling's mouth as well as into the mouths of the chicks she indentifies by sight and recognizes as her own.

Females of a different family of birds in Central and South America have evolved what seems to be a more equitable arrangement with the species they exploit. The giant cowbird *(Scaphidura oryzivorus)* seeks out the large, gourd-shaped hanging nest of an oropendola bird and drops her eggs inside. Once hatched, the cowbirds attack and kill flies that find their way

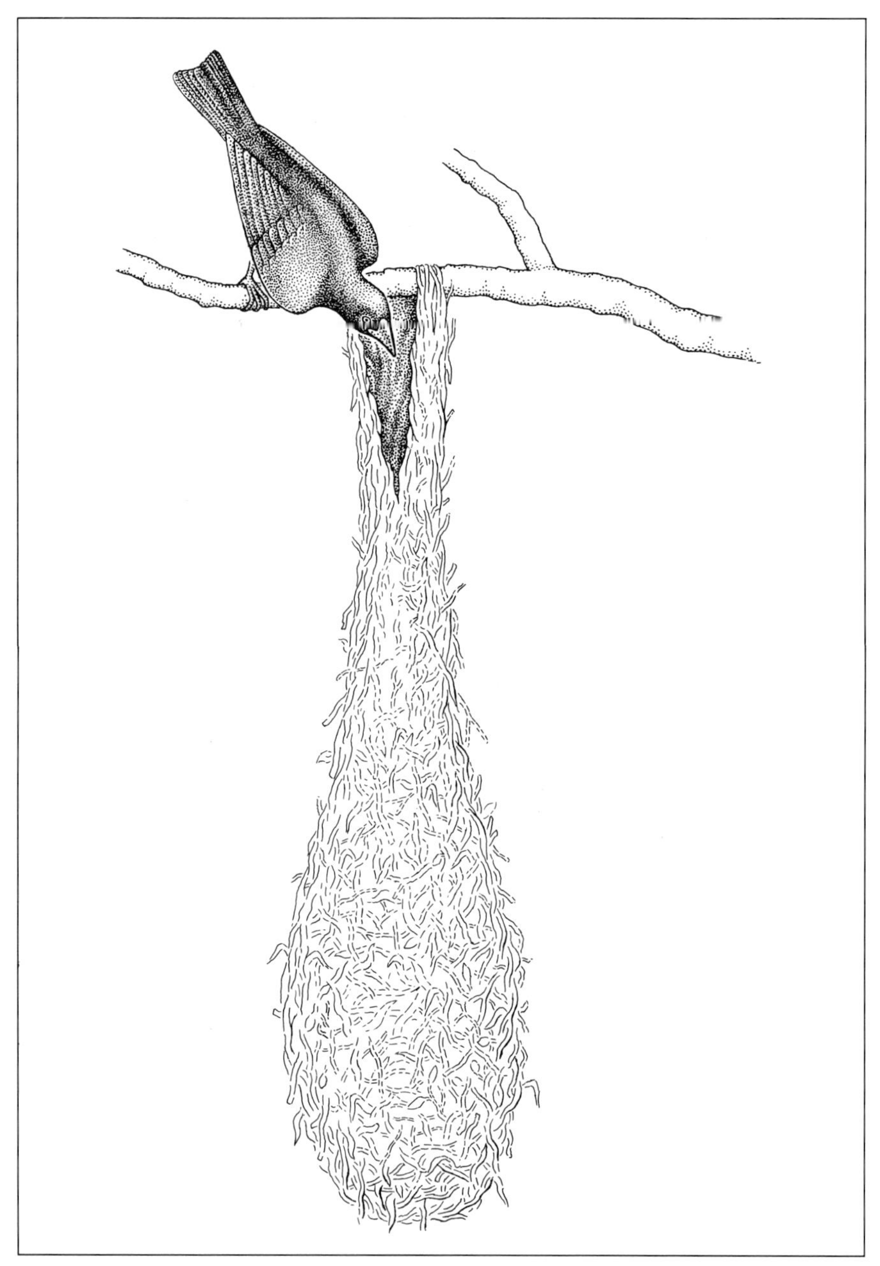

The female cowbird chooses the oropendola's nest as the site for dropping her own eggs.

inside the nest. Moreover, they preen their oropendola nestmates to remove maggots, and thus grow with their broodmates, keeping the nest clean for all of them.

Brood parasitism is not limited to birds, although it is most common there. Fish, such as the Cyprinidae in North America's lakes and streams, seek out the gravel mounds where other fish have built their nests, slip their eggs inside, and leave the host to guard and care for a mixed batch of young, their own and the visitor's, while they swim away to other waters.

Insects as well as vertebrates practice brood parasitism, an indication that this adaptation has evolved independently in many life forms and must be advantageous not only to the parental parasites but to the host species as well. Several species of parasitic wasps lay their eggs with those of host insects. The cuckoo-wasps (Chrysididae) earned their name by laying their eggs in the nests of other wasps. After hatching, the larvae kill and eat the larvae of their host, the mud dauber larvae. In the ant *Formica (microgyna),* the newly fertilized queen seeks a host colony in another ant species. Once she has found a host, she arranges for the adoption of her own larvae by forcibly subduing the workers in the host species and somehow instructing them to turn upon and kill their own queen, after which they take care of the parasite's offspring.

Infanticide and Cannibalism

The kind of mother that tosses her offspring to the wolves, so to speak, or leaves them alone while she concentrates on her "relationship" with her mate, or leaves them on the doorstep of a mother from another species behaves "rationally" in terms of her own evolutionary strategy. Equally understandable in this light is the behavior of those species in which the female willfully destroys some of her own offspring, or does not interfere while another individual commits infanticide.

Evolutionary biologists studying infanticide focus on the intensity of the mother's investment, which is to say, at what moment the mother, or mother-to-be, may be better off trying again. Mother rodents and felines abort fetuses, or resorb the not-yet-implanted embryos, when they apparently receive chemical cues suggesting that the prospective offspring would not survive if it were left to be born. The accumulating evidence from laboratory and field studies of mice makes a convincing case that these animal mothers somehow end these doomed pregnancies before they waste any more energy.

Females apparently abort or resorb fetuses when the social structure in their community makes it likely that the offspring they are carrying would be threatened because of changes in the group. The new factor is the replacement of one male by another, a phenomenon that has been observed in both wild and laboratory mice, in gerbils, and also in captive tree shrews. Observations in equatorial Africa by Brian Bertram seem to verify the occurrence among wild lions of the sacrifice of one set of offspring when new males enter a pride. Shortly after their arrival, the females gave birth in synchrony. Bertram suggests that the new males triggered fetal

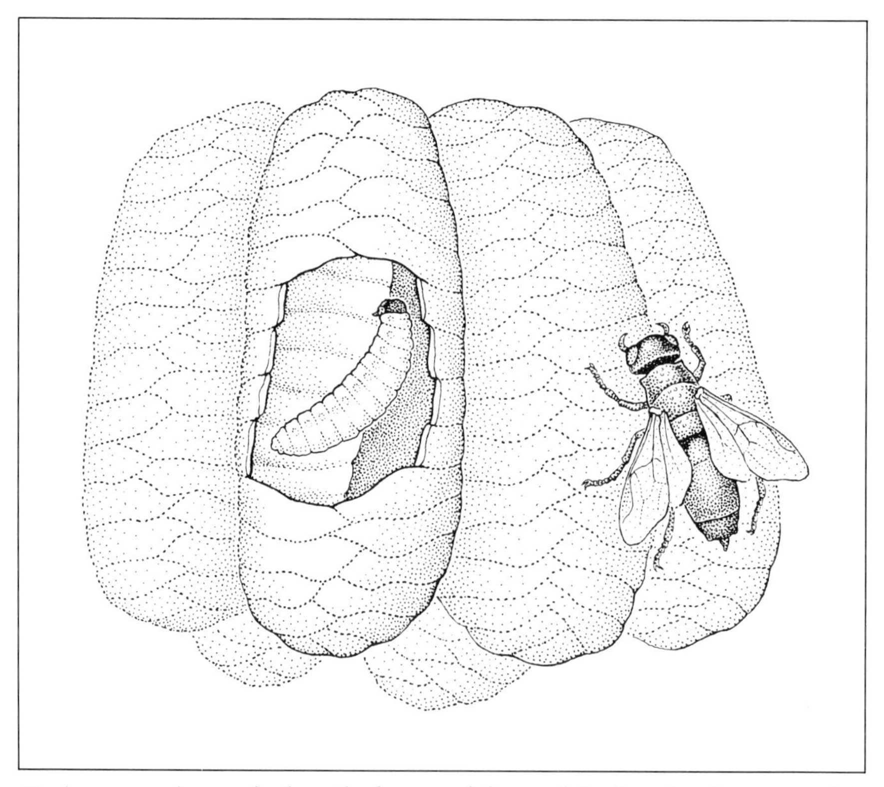

Cuckoo-wasp larvae feed on the larvae of the mud dauber, in whose nest they hatch.

resorption in the females they moved in on, in whom there was no visible sign of miscarriage but some of whom were most likely already pregnant. The new males then reimpregnated them, which accounts for the simultaneity of births.

More dramatic evidence of male-induced abortions has been observed among feral horses in the Granite Range of Nevada. A four-year study of 129 animals unveiled a world of forced copulations and miscarried fetuses. Roaming in bands ranging from four to thirteen mares led by a single stallion, the mares were intermittently taken over by a new male, who forced copulations on many of the females, including some that were already pregnant. All who were less than six months pregnant and were "raped" miscarried. But even those pregnant females not forced to copulate suffered spontaneous abortions, apparently under the stress of the new situation. Most of the females later became pregnant again by the new stallion, confirming the pattern observed among primates and lions. The presence of a new male, although unlooked for by the female in a polygynous group, signals the disposal of the offspring sired by a different male, a kind of prenatal infanticide triggered by the more dominant male and carried out by the pregnant mother.

Zoologists have often speculated about the fate of maimed or unhealthy newborns in the wild. Damaged offspring have been born in captivity, and it is only reasonable to believe that they are born in the wild as well. But they have never been observed there, which leads zoologists to suppose that they must have been done away with posthaste. These hypothetical cases of unwitnessed infanticide can be explained by the same logic as maternal resorption. Such offspring could not survive long and would burden the rest of the group if allowed to live. When primatologist Dian Fossey discovered infant bones in the feces of a gorilla mother whose baby had disappeared, Fossey deduced that the infant must have been maimed or moribund, and thus what looks at first glance like "murder" becomes "mercy killing." With the moribund infant out of the way, the mother will soon stop lactating, start cycling, and be able to become pregnant again.

The killing of apparently healthy, full-term youngsters by their mothers is not common, but infanticide does occur occasionally: witness Ellie the gorilla in the L.A. Zoo, who dismembered her first two babies. She has since delivered several infants by caesarean section that were removed from her immediately and hand-reared. Maternal infanticide has been observed among laboratory-bred marmosets as well. Caged tree-shrew mothers have even devoured their offspring as observers watched. But these events are unusual and probably pathological. The captive mothers involved were under stress and most likely emotionally off-balance when they acted in such a clearly self-destructive way; their behavior cannot be taken as typical of their species.

Robert Martin has studied Belanger's tree shrews *(Tupaia belangeri)* and helped reclassify them from primates to their own order (Scandentia). Noting that caged tree shrews frequently cannibalize their young, Martin describes an elegant experiment at Munich University in which Dietrich Von Holst monitored the effects of stress upon maternal behavior. Von Holst observed that in addition to irregular nursing patterns, the caged mothers secreted an oily substance from a special gland in their chests onto their offspring. This apparently marked them as their own, for when the gland was surgically removed, the mothers ate the young. Von Holst

A lioness resting with her cubs.

then discovered that especially stressed mothers stopped producing this secretion altogether, leaving them, in effect, unable to mark the young for their own safety.

The implications of this discovery apply to wild as well as caged animals. It helps explain the apparently pathological behavior of some captive creatures and may also explain infanticidal practices in the wild, where natural sources of stress, including population pressures, may cause a mother to fail to protect her young. During great shortages of food, maternal infanticide may become a way for the mother simultaneously to eliminate potential competition for food while providing herself with enough protein so that she can survive to try to reproduce again.

Cannibalism—the consuming of the flesh of another creature of the same species—happens often in the wild. In some species the flesh is not consumed until after the animal has died. But other species attack and kill members of their own families on the hoof, and then consume them. Golden hamsters, mice, and guppy fish will turn and eat their neighbor's live newborns and occasionally their own as well.

Most infanticide in nature is committed by males, and often for reasons that make strategic sense from the point of view of the male of the species. Although there is some evidence that mother langurs *(Presbytis)* and pigtail macaques try to defend their youngsters from the assaults of invading males, there is also plenty of evidence that these same females, once their offspring have been killed, immediately solicit sexual relations with the aggressors.

Sarah Hrdy made a thorough study of polygynous troops of Hanuman langurs near Mount Abu, in India, where several females and immatures lived with an adult male leader. When a new male took over a troop, he killed the infants in the group, only to receive sexual solicitations from the "bereaved" mothers. Female primates like these langurs try to defend their young at first, but apparently harbor no feelings of vengeance as a human might, and consort with the new male without glancing backward. They evidently try to become pregnant again as soon as possible. It would seem biologically disadvantageous for the mother to refuse sexual relations after the death of her offspring. Such restraint might have the effect of preventing future infanticides, since aggressive males would not have as many opportunities to pass on their aggressive tendencies; but it would deny the mother another offspring at the same time. Moreover, any female who refused to boycott the aggressive male would be the ultimate beneficiary. Male primates, somehow aware of this evolutionary bind the females are in, appear to kill infants in order to use the infants' mothers to bear their own young.

Infanticide and cannibalism may derive from different strategies to solve different problems. Male infanticide appears to be the deliberate elimination of a competitor's offspring. Cannibalism, on the other hand, seems to be simply a way of filling an empty belly. Hunters have observed male brown bears and polar bears, as well as male pumas, eating cubs whose mothers the hunters themselves had killed. Zoologists do not interpret paternal cannibalism in these instances as infanticide because, without their mothers' milk and care, the infants were doomed. The males' action in killing and eating the cubs, be they the father or just a passerby, is seen simply as making the best of a bad situation.

Males that are not the fathers of murdered offspring are often their killers. But almost as frequently the killers are females in competition with the mother over limited resources—food, warmth, or sexual partners. In some species of lemmings—the tiny rodent that inhabits Scandinavia and the northern reaches of North America—newborn pups are eaten by

In captivity, the female tree shrew sometimes cannibalizes her own young.

almost everyone *except* the mother and the father. Something about constant contact, most likely smell, protects the newborn from becoming a parental meal. Among laboratory lemmings, when the father lives with the mother throughout her pregnancy, he may even become a helper, bringing food and warming the pups when they arrive. But if he is removed from her cage and is not continually receiving the chemical messages she sends out, on reentry he might resort to killing and eating his offspring just as a strange male would.

In most species, living with the newborn triggers protective behavior in both males and females. When biologists are investigating laboratory animals like rodents, they can sample hormone levels in the blood that reflect the degree of tension in the adult animals toward the infants, and toward one another. In a laboratory experiment with collared lemmings, Canadian zoologists divided females with their litters into three groups. The control group consisted of just mothers and offspring, the second group included males with whom the females had once mated, and the last group was made up of strange males. The first two groups remained stable, but the strange males almost immediately killed 42 percent of the pups they found. Moreover, the females in that last group produced litters sired by the strange males in just three weeks, while the females whose previous litters were not killed by unfamiliar males gave birth again in four weeks. A week is a substantial amount of time in the life of Arctic creatures with a short breeding season, and the advantages of an extra week of life for the offspring before winter makes reproductive success more likely for a male who has disposed of the female's earlier litter in order to hasten the birth of his own offspring

Evidence of maternal infanticide is sparse. With so many species in the wild, and so few research projects completed, it is hazardous to generalize. Even in careful investigations the nature of the research makes foggy patches inevitable. With red howler monkeys, for instance, there has been a good deal of observation recording the disappearance of very small infants. But how they disappeared—whether they died because they were weak, sick, hungry, or were abandoned or murdered by mother or father—is guesswork.

What seems fair to conclude is that mothers, fathers, and siblings have very different interests when it comes to the health and development of infants of most species. Mothers may have to relinquish their young to the demands of stronger males. But the same langur who mated with the male who killed her infant may have fought hard to defend that infant's life. In the animal world, among those females that take on parental responsibilities for whatever reason, mothers protect and nurture their offspring. Most females are "responsible" mothers—they look after the well-being of their young. But what that means varies enormously according to the species.

Favoritism

Some mothers may indeed sacrifice their own well-being for offspring, but they do not necessarily treat all of their offspring equally. In fact, equal treatment is only really apparent in the wild in cases of omission. Among species with no parental care, each offspring may be said to be neglected to the same degree. But in many species, mothers play favorites. They may feed an older youngster at the expense of a younger sibling, and even allow the elder to devour the younger. They may favor offspring of one sex over the other, giving weight to the argument that females of the species are singlemindedly intent upon leaving a successor. They achieve this end by behaving in a variety of ways that seem to have been evolutionarily successful.

For polygynous mammals, those in which many females share a single male, lifetime reproductive success varies greatly for males, and to a much smaller degree for females. Those males a mother produces have a much better chance at reproduction if they are larger and stronger than their competitors. Among the red deer *(Cervus elaphus)* on the Isle of Rhum in Scotland, the male infants are born larger than females and they proceed to extract a greater quantity of milk while nursing, so that the males grow faster and larger than the females. They demand so much from their mothers, in fact, that the mother of a male is likely to forego pregnancy the next year, while mothers of females continue to produce a new infant annually. Female deer could be said to favor individual male offspring above individual females because they nurse them longer and allow them to take more milk. However, males leave their mothers' sides earlier than females; thus a mother's lifetime dedication to offspring of both sexes may be equal in the long run, just crowded into a smaller span of time for sons.

Or it may be that a mother can expect a greater reproductive pay-off by producing a son who will dominate many females for enough time to greatly enhance her genetic contribution into the next generation. More female deer seem to be born than males, and this argues against male favoritism.

Female offspring, on the other hand, receive the most attention among a large group of insects. In colonial or social bees, the workers, all of whom are female, give larger quantities of food to the female nymphs that they are taking care of in their hexagonal nursery cells. The male bees that will eventually fertilize another queen receive scant attention. Perhaps this is because many fewer of them will be needed as adults, or perhaps because they are not as closely related to their sister-nannies and so are less worthwhile from a passing-on-of-the-genes point of view. Whatever the reason, male bees fare poorly as youngsters, and as adults live only long enough to pass on their sperm and die.

Sometimes favoritism operates at the embryo stage in the form of egg cannibalism. Among parasitic wasps, the female lays a lot of eggs, and those that hatch first turn to their unhatched potential siblings as food. Termites, wasps, and bees all engage in some form of egg cannibalism. Newly mated termite queens and kings routinely eat part of their first batch of eggs and larvae. Workers in some species of bees lay trophic eggs, eggs immediately eaten by the queen and larvae; these eggs are evidently produced only as a food source for the developing offspring. In some species of ants, the trophic eggs fuse into a mass that entymologists have labeled "trophic omelettes." Other insects produce another kind of nonreproductive egg. Instead of providing food for its egg-siblings, the eggs of Mexican owlflies provide a line of defense. The females lay their real eggs in packets on the sides of twigs, then deposit a group of modified eggs further down the stem that function as a kind of quicksand, stopping predators from approaching the true eggs that are holding the next generation.

The phenomenon of producing eggs, or even more developed organisms like embryos, whose only purpose is to feed or protect the surviving offspring appears either wasteful or extremely clever, depending on the value placed on "life" and the value placed on perpetuating an individual animal. When one embryo feeds off its littermates, as happens in several species of shark, the practice is called *fetal cannibalism,* or more literarily *cainism,* after the Biblical fratricide. An ichthyologist, Stewart Springer, made this discovery on the dissection table where, while exploring the interior of a large sand shark, he was nipped on the hand by a living nine-inch pup inside its mother. Further investigation revealed that the shark's pregnancy begins with a dozen or more fetuses that eat one another inside the oviduct until only a single sand shark survives to be born. This prenatal cannibalism explains, in part, why sharks in the several species that practice this prenatal rivalry are born with functioning teeth: they have been using them as fetuses to kill and eat their unborn siblings.

Among many birds of prey, older siblings eat their younger nest mates. This form of "siblicide" has been occasionally observed among eagles. The female sometimes lays two eggs several days apart, and the older eaglet apparently awaits the birth of its brother or sister, only to make it into a meal within a few days. Some biologists label this situation parental infanticide on the supposition that the mother deliberately encourages the older, stronger offspring to feast off its younger sibling.

eleven

Enlisting Male Support

MOTHER AND INFANT are a familiar duo, but that is not the whole story. Few mothers and offspring in social species live isolated from mates and, above all, from other females. The animal world abounds in social units—troops and broods, herds and hives complete with males as well as females, and cousins, brothers, sisters, and aunts along with fecund mothers. Although childbearing or brooding, and later on care and coaching, fall heavily on the mother in most species, social units larger than mother and child provide both mothers and young with assistance. Males, especially when they are the fathers, may help to defend and feed new offspring. In many species, siblings pitch in when a new youngster arrives. Most mothers accept the aid and use the time away from dependent offspring to find food for themselves and to perform other tasks that they cannot do while burdened with an infant.

Males of the species have a vested interest in caring for their offspring from the moment of conception when they can be reasonably certain that the new life in question is their own. How they ascertain paternity is largely unknown. But they must have some way of judging because males in so many species do help, and selectively.

Within the fish families that have been studied, about a third of the parents give some kind of care, and about half the care is performed by the male alone. Some guard the territory where the eggs wait in a nest, and others carry the eggs around with them—in their mouths, if they are mouth-brooders, or in a variety of other peculiar places including an attachment to the lower lip in a species of Central American catfish *(Lori-*

vincing example of a chemical partnership between parents. Female cockroaches release very little uric acid, but males of this species store urates in specialized cells or secrete them into accessory glands. As soon as the cockroaches finish copulating, the male widens his genital chamber to expose a white secretion of urates. The female, sexually satisfied, then eats the urates, which she needs to feed her eggs because she cannot manufacture uric acid. Those female cockroaches that ate uric acid after mating laid their eggs almost immediately, while those that for some reason did not find this source of food usually resorbed the eggs, or remated, or both. It seems as though the successful male cockroaches are those that can accumulate a lot of urates, which ensure a healthy brood.

Recent laboratory experiments with katydids, the musical orthoptera of the United States, have indicated that, given a choice between two singing, courting males, female katydids always choose the one with the largest spermatophore—the sperm-containing packet—that can provide the best nourishment for their fertilized eggs. Like the cockroach, the female katydid has no access to this food without a male. Complementary experiments with the Australian bushcricket *(Tettigoniidae)* involving radioactive labeling have revealed that nutrients derived from the spermatophore are incorporated into the eggs. With the bushcrickets, the females actively preferred males with very large spermatophores, which they, too, ate immediately after mating. Those that ate the largest spermatophores produced both more and larger eggs, which in insects correlates with healthier offspring. The researchers point out that the male sacrifices for his offspring when he produces an especially large spermatophore, which represents up to 40 percent of his body weight. But this large spermatophore enhances his appeal to the female of his species and accounts for his selection as a father in the first place.

In some polygynous baboon families, a single male guards his entire harem and the offspring, protecting them all from enemies within his species and without. The female seems to depend on her mother and sisters for some kinds of aid, but it is to her male protector that she looks for defense. Among chacma baboons, the alpha male (the leader of the troop) apparently seeks out mother–infant pairs and protects them from aggressors by making threatening gestures with his enlarged canines as well as by actual fighting. In response, infant baboons seek out the adult male rather than their mother when they feel endangered. Male primates frequently play defender.

But in other species males seem to function as more general, all-round helpers. One example is the male *Lamprologus brichardi,* the six-and-a-half-centimeter gray cichlid fish of Lake Tanganyika. In the summer of 1978, researchers from the Max Planck Institute dived into the lake from the Barundi shore and, with delicate skill, marked two hundred individual fish by injecting blue dye into their scales. These fish live in family groups that include parents and young of different sizes. They are egg-laying and pass through a larval stage before becoming fry. The parents enlist their younger offspring in territorial protection as well as the job of removing predatory snails and other particles from the eggs and larvae. The investigators observed 27 fish families during four weeks of diving and noted that the offspring stayed together even after the parents had been removed by the scientists. But as soon as the young reached five and a half centimeters, the young males left their parents to join a bachelor group. Eventually some of the males found mates and territories and started families of their own. Others never did. Still a puzzle to zoologists is why these fish remain with their families so long and help care for the brood. It could be that the individuals benefit when their kin benefit, or a kind of self-interest may be at work in which each fish hangs around in exchange for the territory upon the death of the parents.

Male mammals and birds, unlike some male fish and reptiles, can never be absolutely certain that the new generation they are protecting is genetically their own. A number of evolutionary biologists argue that the degree of paternal concern appears to be directly proportional to the degree of certainty males can have about their paternity. This hypothesis makes sense for those animals that use external fertilization. It certainly explains why cichlid fathers such as the mouth-brooders should be willing to starve themselves to protect the young. It does not really explain the mother's readiness to abandon her charges, since she is no less sure of her maternity simply because the male is more certain of his paternity.

Perhaps the care of youngsters really is an onerous chore which neither parent relishes and both try to evade. Nurture then falls to the parent least able to cut and run, the parent left with the infant or the clutch of eggs. This, in fact, is the theory advanced by Richard Dawkins, who has proposed that natural selection favors the desertion of offspring by the first parent that gets the chance. He supports this theory with the obvious example of internally fertilized females, such as mammals, that are forced into a "cruel bind" of providing care because they are obliged to carry developing embryos, while their inseminators may or may not disappear. At the opposite pole Dawkins places externally fertilized species such as many fish, where females spawn first and flee and the males fertilize and then remain to tend their eggs. This theory works better on paper than in the pond. Overall, the method of fertilization does determine which parent cares for the offspring—internally fertilized vertebrates are cared for by females and externally fertilized offspring by males. But males in internally fertilized species do not always desert when they have the opportunity. And in externally fertilized fish and amphibians, parental care seems to correlate with male territoriality. Males that are territorial tend to provide for their offspring, because they can be fairly certain that the majority are their own.

Child Care in Monogamous Species

Among monogamous species of mammals, parents are close to the same size and often look so much alike that it is hard for observers to distinguish them from each other. Size implies strength, so it is not surprising that monogamous males carry only about 50 percent of the responsibility for defending the family when the female is as strong, or stronger. What emerges within these species is a high degree of role sharing and job sharing. Looking at it the other way around, neither parent is so specialized, except for pregnant or nursing mothers, that their jobs cannot be interchanged.

An instance of one of these highly cooperative types is the marmoset. This small New World monkey of tropical South American rain forests has been a favorite among Europeans since its discovery by the Spanish conquistadores. The furry primates fit comfortably into the collars and cuffs of the ladies of the Spanish court and became a cross between a fashion accessory and a lap pet for the next two centuries. Despite their popularity at court, very little was known about marmoset behavior in the wild until recently because of their remote habitat. They have been studied extensively in the laboratory, and some new field studies now complement the laboratory work. What emerges is the picture of a family unit that includes a breeding female, one or more males, and juvenile offspring, often in sets of surviving twins. When the young appear after 148 days of gestation, their collective body weight is almost one-third that of their mother's. The mother feeds herself and suckles the infants for a day or as long as two weeks, depending on the species, until the males take charge, carrying the youngsters about and bringing them to their mother when it is time to nurse. Both parents share the care of their first set of offspring equally. But with the arrival of a new set of twins, the older pair do much of the child care, while the mother uses most of her time to eat as much as possible in order to produce enough milk to sustain her growing family.

Patricia Wright followed titi monkeys *(Callicebus moloch)* and owl monkeys *(Aotus trivirgatus)* at the Cocha Cashu Biological Station in Manu National Park in Peru for fifteen months, and found that titi fathers carried their infants 92 percent of the time. And while the mother may have nursed the babies, it was the father that taught them how to eat solid foods. In fact, the mother rarely shared food with her infants, while the father willingly offered them fruit and insects throughout their first year of life. Morever, the father played with the offspring, leaving the mother to lead the group while he trailed behind carrying the growing infant.

Cooperative rearing of offspring is common among canids such as foxes, wolves, coyotes, and jackals, animals that tend to live monogamously. Female dogs frequently remain in the burrow that canines usually seek as temporary homes for their new offspring; the males leave to hunt and bring back freshly killed meat, which more often than not they regurgitate, partially digested, as food for the pups. African blackbacked jackals live in long-term, sometimes lifelong, pairs and enjoin one generation of offspring to help care for the next. Zoologists watched fourteen pairs of the small omnivores for over three and a half years on the Serengeti Plain and noted that males and females shared responsibilities about evenly. They groomed each other, hunted cooperatively, and guarded their pups during the youngsters' early weeks. While the female provided milk, the male regurgitated food. As the season progressed and additional litters arrived, the older siblings helped provision the youngsters. Both helpers and new

In owl monkeys, as well as titi monkeys, the father cares for the youngsters most of the time.

arrivals benefit from this arrangement. Fresh meat determines survival, and the more they bring in, the better the chances for all of them.

Monogamous birds, unlike most paired mammals, are not always alike in size; predators like female hawks and owls often grow to twice the bulk of their consorts. But as partners, these birds vie in their concern for their unhatched progeny. This has proved almost fatal to the California condor *(Gymnogyps californianus),* which verges on extinction. Competitive parenting within a condor couple was witnessed by wildlife biologists in March 1982. As they watched through telescopes, the parent birds pushed and fought with each other in their montane nest, competing so fiercely for incubation rights that the egg accidentally rolled out of the nest to disaster on a ledge far below.

More frequently, monogamous species are monomorphic and divide parental responsibilities equally. Among birds, of course, the female is not tied to the nest by an obligation to feed her offspring from her own body. Thus, as with fairy penguins that live near Melbourne, Australia, on Phillip Island, male and female alternate daily egg-sitting so the other can look for food. After the chicks hatch, the parent birds continue this pattern, one babysitting while the other gets a food supply, the sitter keeping the chicks warm beneath his or her feathers or inside the breeding burrows that the parents had prepared for their offspring some months earlier. The great blue heron *(Ardea herodias)* and the common potoo (Nyctibiidae) also split their brooding equally, the one resting on the nest for several hours at a time while the other searches for food, then switching tasks after a while. Doves (Columbidae) divide their responsibilities more rigidly, the females incubating the eggs all night, the male relieving her at dawn and holding the roost until afternoon.

In other birds such as the California roadrunner *(Geococcx californianus)* and the spotted sandpiper, the female sits on the clutch part of the day, the male all night. Recent studies reveal that one of the ways roadrunners economize on energy is to reduce their body temperature at night. Females that are laying eggs need especially to conserve energy. Thermometers implanted in the birds revealed that all females and nonmated males lowered their temperatures after dark, but the mated males, those seated on the clutches, maintained the steady daytime body heat necessary to keep the eggs alive.

Waterfowl such as black and white swans (Anatidae) take turns incubating their eggs, with the female emitting a bugle-like signal at the changing of the guard. After hatching, the cygnets follow both parents about the surface of the lake, and both parents rush to the rescue if a cygnet is in trouble.

Joint child care is the rule rather than the exception among animals that live in pairs for the greater part of their adult lives. But male birds are far more involved than male mammals with their offspring, since they can do everything for the young except lay eggs, while the most nurturing male mammal can neither gestate nor suckle his young. Male birds behave as if

A roadrunner incubating eggs.

they have a security as to their paternity that is perhaps unjustified. Whatever the reason, more birds than mammals live in small family units; and the extent to which a male helps defend, feed, and teach the young may depend as much on the social system within which he lives as on his absolute certainty that the young are his.

Danger is always present in the wild, and occasionally when predators threaten a mother with her young, the mother alone succumbs. This can signal the death-knell for the orphan as well if that youngster is still dependent for food or protection. Among monogamous birds where the parents have been jointly sharing child care, it is common for the father to continue bringing back food to the nest and to help prepare the young for flight. If the mother is killed, the offspring have a smaller chance of survival because they must be left alone for a part of the time when the father is out foraging. But a surprising number of motherless birds survive, among them willow ptarmigans, cousins of the common chicken. In controlled studies, father ptarmigans were left to raise their chicks and did so successfully, showing them what to eat by day and guarding and huddling close to warm them at night. In contrast, death came to those chicks in the control group that were left completely without parental care.

Among the African apes, whether an orphan that is no longer nursing will survive or not seems to depend on the age of the orphan as well as on the temperaments of the individuals involved. A photograph from Rwanda shows a very young female gorilla, dubbed Papoose by her human observers, being carried by a huge silverback male. The male may or may not have been her father. But he led the troup her mother had joined, and after the mother's disappearance the old male traveled with the feisty infant on his back, letting her munch wild celery alongside him during their daylight forages and share his nest beneath the trees at night.

Papoose's story is not unique. Observers have witnessed equally compassionate behavior among chimpanzees and baboons who, at some sacrifice to their own well-being, have cared for orphaned youngsters. The ability to nurture may be stronger among females, especially among mammals where the young must nurse, but it is not an exclusive talent. Among group-living species, familiarity seems to breed something akin to compassion. At the least, the life-saving acts of male primates toward orphans suggest that in some way the males find such actions rewarding.

part iv

Sisterhood

twelve

Working Together

WHETHER they are sniffing about in the underbrush, sopping up sunlight on a dry rock, or avoiding the glow of a harvest moon, many animals spend much of their lives interacting with others of their species. These social creatures live together because the advantages of cooperation apparently outweigh those of solitude.

Of course there are loners—sea cucumbers on the ocean floor, grasshoppers, for which the act of fertilization is their only social experience. Solitary creatures struggle against the environment, neither helping nor hindering each other to any degree. Not so those species that cluster in a diverse array of aggregations and communities based largely on some degree of kinship—be it group, herd, hive, pride, or colony—that may endure a few hours, days, or years, the entire lifetimes of the individuals involved, and sometimes for generations.

Among monogamous species such as beavers, half of all the South American primates, including some marmosets and titi monkeys, some cichlid fish, and most colonial birds and wolves, the social group is the nuclear family—mother, father, and offspring. Here both parents tend to split chores equally. The young males in these mammal species often leave the nuclear unit early to find territories of their own, but the females usually remain, helping care for their siblings as, season after season, new young increase the family.

But in many species males, including fathers, bear little or no responsibility for care of the young. It is not surprising, then, to find cooperation among the females. Within elephant and lion groups there are bonds between females that last for years. Although each adult female may copulate with half a dozen males in the course of her life, males do not take up very much of her time. When not in the seasonal rut, males of these species wander off to lives of bachelor solitude. But the females remain together. The responsibilities of incipient motherhood and then the actual care of the young seem to create a mutual dependency. Childbearing and rearing for these females is not a one-shot operation. The immature insects, fish, birds, and mammals in social communities need to be fed and protected for a long time after they are born or hatched long, that is, in proportion to their mothers' lives.

Bottlenosed dolphins and elephant mothers find giving birth itself to be a difficult process, too difficult to do alone. The pregnant dolphin enlists the help of another female—whether related or not is still unknown—early in her pregnancy, and that midwife dolphin stays close to her and is with her at the delivery. The pup emerges, tail first, which prevents the infant from strangling on water before it is fully born and able to control its lungs. As soon as its head emerges, both midwife and mother nudge it to the surface where the twenty-five pound newborn begins to breath. Eventually the pup has to submerge to find its mother's nipples. In comparison, the elephant helper has an easier task. The elephant midwife assists by pulling the 200-pound baby from its mother's uterus, where it has been growing for almost eighteen months.

Young elephant calves of both sexes act as herders when their group is on the move, making sure the littlest elephant is not left behind. In some instances the adolescent female may even suckle the youngster she is guarding, lactating despite the fact that she has not yet given birth herself. Feeding and herding her younger brothers and sisters, young elephants are well prepared for motherhood when the time comes.

Most female elephants in a given location come into estrus at about the same time, and most probably mate with the same male. This means that all young elephants of the same generation in a single group are likely to be half-siblings sharing a father. The help they give the young is thus help within the family. This high degree of kinship makes it more reasonable in terms of the individual's self-interest for elephant mothers to allow their neighbor's offspring to suckle alongside their own. A theory known as *inclusive fitness* or *kin selection* has been put forward to explain how this self-interest works.

Mother and "auntie" bottle-nosed dolphins with a calf.

Inclusive Fitness

Increasingly, behavioral biologists are finding evidence that individuals within many species—insects as well as mammals—recognize their own close relatives and in some cases behave generously toward one another in proportion to the degree of relatedness.

Gary McCracken, working among Mexican free-tailed bats *(Tadarida brasiliensis)* in caves in Texas, has discovered an uncanny ability of mothers to identify their own young amid the pups that are densely clustered on the walls and ceilings. Females give birth to one pup at a time, which they deposit in the crowd of newborns within hours, returning twice a day to nurse them. The intense clustering seems important to stabilize the infants' body temperatures. In a particularly elegant experiment, researchers picked pairs of nursing mothers and infants and used chemically identifiable genetic markers to determine the relationship between mother and pup. Although it was difficult to distinguish pairs in the dense "creche," they selected pairs as close to the center as possible. The

Infant Mexican free-tailed bats crowd together inside a cave.

blood samples of both animals, as well as small pieces of muscle tissue from the mother, disclosed that mother bats found and nursed their own pups in the vast crowd 83 percent of the time. Only 17 percent of the mothers appeared to be feeding genetic strangers. What is unknown is *how* the mothers were able to identify their own offspring within the densely packed horde of hungry infants.

Sherman's work with Belding's ground squirrels in California indicates that the mother squirrels identify their offspring by smell. Using electrophoresis to determine the relatedness of individuals, Sherman found that not only can mothers identify their young, but female siblings can recognize both whole and half sisters. As adults, these females volunteer to assist one another in proportion to their relatedness. Among ground squirrels, the crucial period for learning relatedness occurs at the age of three weeks, when pups make their first foray above ground, the same time that mother–offspring recognition occurs. Whereas in ground squirrels the olfactory cue seems to be the key to mother–offspring recognition, the ability of siblings to distinguish degrees of relatedness seems to have a different source.

The recognition of close relatives by individuals who have never seen each other before has been called *phenotypic matching*—an ability to be aware of one's own appearance and somehow compare it with someone else's so that the individual matches up the "stranger" and knows her as a "sister." Once relatedness has been established, a variety of cooperative behaviors occur, depending on the species, to ensure that the genome the two females to some extent share will survive. In nature, eusocial sweat bees work together in units of up to fourteen degrees of kinship. When raised in solitary confinement in the laboratory and then released together at maturity, the bees immediately congregated in groups according to their degrees of relatedness. Without ever having met each other before, these bees identified and preferred their closest kin.

In a refinement of Darwin's concept of fitness, William D. Hamilton and Robert Trivers used mathematical models to explain not only why females in many species cooperate with their sisters but why some females leave no offspring at all. In social ants, bees, and wasps, unfertilized eggs develop into males and fertilized eggs into females. The males are *haploid,* which means that they have only half a set of chromosomes. They have inherited this genetic endowment from their mothers. Most higher organisms have a full set, having received a half set from each parent. Organisms with this type of genetic endowment are called *diploid.* In ants, bees, and wasps the females, unlike the males, have developed from fertilized eggs; therefore they have inherited half of their genes from their mother, the rest from their father, and are diploid. But because sisters all inherit the same genes from their father, since he has only one set to pass along, 50 percent of the sisters' genes will be identical. The other half will come from their mother, and since she can give them 50 percent of her genetic endowment, they will on average have half of their maternal genes in common. Thus, on the average, sisters will be 75 percent related to each other — 50 percent from having the same father and 25 percent from having the same mother. In higher animals where both parents are diploid, siblings average only 50

Infant Belding's ground squirrels at three weeks of age, when they establish their maternal bond and kin recognition.

percent of genes in common. Communities in which the males are haploid and the females diploid are called *haplodiploid.*

The rationale given to explain the behavior of sterile females in this situation is that the female siblings have a higher proportion of genes in common with each other than they would with their own offspring. So it makes genetic sense for them to remain sterile and to help their mother produce more sisters.

When the preferential treatment of an animal toward its kin increases the chances that the genes they share in common will survive into the next generation, the animals are said to be enhancing their *inclusive fitness* through the process of *kin selection.* An example of inclusive fitness and kin selection at work can be found among the paper wasps *(Polistes).* This species earned its name from the masticated wood pulp, which looks like paper, that it manufactures and uses to make its hives. Some females of this species lay their eggs alone in a nest and raise the larvae themselves. Others share nests, usually with a sister, who cooperates in feeding and caring for her nieces and nephews. Although the helping sister remains childless, the cooperative nests produce more healthy young wasps than those of other paper wasps nesting alone. The pleasures of sociality aside, the cooperative nest seems to be evolutionarily more advantageous than the single venture. There might be a time in the evolutionary future when all paper wasps nest communally as the cooperative individuals gradually outnumber the loners.

The helping sister, the maiden aunt of the wasp world, would probably not devote her energy to a stranger's offspring. With a few interesting exceptions like some warrior ants, cooperation among females in the animal world depends on kinship ties. Females will help their relatives where they would not help strangers. The specific way in which kin selection operates in these species is called *nepotism,* an unfortunate term for female behavior, given its derivation from a root meaning "nephew," the preferred favorites of certain political figures in the Renaissance. But for the moment, nepotism will have to mean *niecism* and refer to the predisposition of female animals to help their kin.

Alloparenting

The term *auntie* is British and in animal behavior refers to a helper, not necessarily an aunt, who aids the mother. She may be a younger animal that is not yet sexually mature, or a contemporary that for any of a number of reasons does not produce her own offspring. The role has been observed repeatedly among primates such as chimpanzees, lemurs, baboons, and macaques, but not among groups of gorillas or orangutans. *Alloparenting* (as this aunting process is also called) only occurs within complicated societies such as the social insects, birds, or mammals, in which females that may not reproduce devote their lives instead to caring for the offspring of their mother or sisters. The hazards of relinquishing offspring to an alloparent are obvious, especially among primates where more than simple reflexes or instinct motivates each animal. The advantages are clearly also rewarding. The virgin female gets some good experience, the mother gets a chance to forage away from her offspring, and if anything happens to her, there is a good chance that her offspring will be raised and nurtured to maturity anyway.

In studies among anubis baboons *(Papio anubis)* in Kenya, Linda Scott observed the heretofore overlooked role of adolescent females. Whereas male monkeys had been described as having an extended adolescence, females had been thought to jump immediately from childhood to adulthood at the first estrus. However, Scott notes that these female primates usually experience at least eight estrous cycles before they become pregnant, a period during which they often become alloparents, learning by doing appropriate maternal behavior.

William Hamilton, Curt Busse, and Kenneth S. Smith studied chacma baboons *(Papio ursinus)* in Botswana and reported on the adoption of nine of ten natural orphans by four- and five-year-old adolescent baboons, as well as the adoption of two orphans experimentally introduced into the community. The researchers concur that the one- to two-year interval between first estrus and first pregnancy serves as a time for the young females to gain parenting experience by adopting, when they have little to lose in the experiment.

Few doubt that alloparenting is a learning experience for the adolescent, but there is some controversy over the adaptive benefits of alloparenting to the borrowed youngster or its mother. A ten-year study of captive langurs by Phyllis Dolhinow revealed that the only infants who received attention had healthy, caring mothers, and thus did not really need the extra help. On the contrary, orphans and rejected infants were ignored by other females in the group. Thus, in langurs, the alloparent helped herself, but healthy orphans that might have survived were ignored, thus depriving the community of a larger number of possible members.

The helping female is not always young, and the help she gives is not always motivated by immediate self-interest. In fact zoologists have noted that within some mammal groups, the older, postreproductive females play a vital role, one that sometimes jeopardizes their own lives. Among

An adult female Belding's ground squirrel sounds a warning, at some risk to herself.

Belding's ground squirrels, for instance, a female sentry sounds an alarm of chirps that draws attention to herself while warning the others of territorial predators; she is usually an older sister or a grandmother that has already produced several litters. Observers have noted that more callers than noncallers do, indeed, succumb to these predators themselves, which make them good examples of nature's "altruists."

Sarah Hrdy observed older females among groups of langurs acting as lookouts on the fringes of their territory in a rather sophisticated form of group defense. Older rabbits and deer raise their tails like flags, the white underside a warning to the young to run and hide, even as it signals to the hunter that they are there. And the gazelle, with no such pigmented tail, jumps high with stiffened legs, making herself the target even as she warns her young.

The older females can afford to take risks as they are no longer responsible for dependent young. Their acts can be seen as repayment for risks that were taken for them by an earlier generation. This kind of behavior is identified as *reciprocal altruism,* an apparently selfless act which is not so selfless after all.

Yet there is not always an obvious quid pro quo explanation for all apparently selfless behavior. Postreproductive lionesses with worn down or missing teeth are able to survive for twenty years or more because, as part of the pride, they are never ostracized but are cared for by younger females who do their hunting for them. Females of some species develop bonds among themselves that do not have an immediate reproductive benefit. This behavior apparently fills a different kind of social or biological need.

Postreproductive mammals do not experience a physiological change akin to human menopause, at least not one that has been verified anatomically. Yet among large species especially, notably elephants and pilot whales, females that have not produced offspring for over a decade are leaders of their groups. These elderly elephants have been found with milk in their teats years after they last gave birth. And at their deaths, what seems like a mourning process has been observed, as the other elephants milled around near the corpse and then buried it under piles of leaves and brush. Reports on pilot whales indicate that postreproductive females make up as much as a quarter of any pod. And they too continue lactating for as long as twenty years after their last delivery. Corpses of whales as old as seventeen with milk in their stomachs have been picked up in fisherman's nets. These animals are clearly living on other foods and probably suckle as a way of maintaining either a bond with their mothers, or bonds within the group as a whole.

Adoptions

When the old chimpanzee Flo was dying at Gombe, she held on to her youngest son, Flint, perhaps because without a younger offspring she did not have any way of releasing the child from her care. After Flo died, Flint's sister, Fifi, visited him and encouraged him to join her in the forest. But the eight-year-old Flint was too emotionally crippled to survive on his own.

Flint's demise is one of a handful of juvenile chimpanzee deaths witnessed by scientists. It is probably unusual only in that it was witnessed. But in the wild, according to extrapolations from observations, at least half of all mammal young succumb within their first year. Throughout the animal kingdom the majority of those offspring that have managed to be born probably do not make it to maturity. These rates are strikingly out of kilter with the survival rates of animals born in captivity. Horse breeders, for instance, would not accept the probability that at least half their foals would never have a second birthday. Likewise, zoo directors struggle to maintain every infant gorilla, resorting to incubation, bottle feeding, and occasionally caesarean deliveries.

In nature, mothers fare better than their offspring. While the young frequently succumb to predators and illness in their first year, adult females usually mate again and produce more infants or broods the next time around. It is far more frequent in the wild to find mothers bereft of infants than the reverse.

But occasionally the mother of altricial young, which cannot survive without intense care, disappears. The young orphan or orphans she leaves behind will only survive if another female adopts them. Small mammals still dependent on their mother's milk would seem to be in an especially desperate predicament. Yet enough of them have survived to make adoption, or stepmothering, by other females something to be taken into account in any description of female behavior.

But what appear at first glance to be adoptions may be in reality kidnappings. Junichirô Itani and Thelma Rowell observed bereaved mother macaques not only accepting other infants at their breasts but on occasion actually stealing babies from their own mothers. If the stepmother has milk of her own, the baby is usually all right. Occasionally, however, the kidnapper is unable to nurse the infant, and if the real mother fails to retrieve it, the baby dies.

The kidnapping of a helpless infant highlights the apparently strong urge of some female animals to go through all the motions of motherhood once the process has started with pregnancy and then birth. These individuals become desperate at the death of a baby and appear to try to offset their loss. It seems as if they need a baby to satisfy a biological urge. What enables females of these species to become stepmothers is not generosity so much as a delayed development of maternal attachment to an individual offspring.

This lack of early discrimination among some bereft mothers is evolutionarily beneficial to the species. It provides nurture for orphaned animals who might otherwise not survive. The female macacque that holds just any small macaque is not caring for a particular child, but in an emergency her behavior would help that small monkey to live.

In other species, however, the same mechanisms that enable a mother elephant seal to identify her own offspring's squeals and smells tend to make her avoid and certainly not seek to help the offspring of others of her species. The elephant seals *Mirounga angustirostris* emerge from the sea yearly on Anno Nuevo island off the coast of California and meet in enormous rookeries of literally hundreds of animals, all about to give birth. Females will very occasionally nurse an orphaned pup alongside a pup of their own, but this ostensibly generous act is usually detrimental to both youngsters. Each pup requires an enormous amount of rich milk to build up a thick supply of fat for insulation during the long winter at sea, and no mother seal has enough milk to provide for two pups. The mother is in reality sacrificing the health and survival of her own offspring as well as that of the orphan whose life she has prolonged but not really saved.

But the situation is different when the mother has lost a pup. Many seal pups die before they can reach the sea, the victims of adult females who sometimes bite them to death or simply roll their gigantic bodies over them until they smother. Yet while some infant seals die because of the clumsiness of their own, or neighbor, mothers, a number of orphan pups are saved by these bereft foster mothers, who feed them until they are strong enough to go off on their own. A female will only come into estrus if she nurses a pup and is thus under strong biological pressure to find a suckling infant.

The feeding of a stranger's offspring, and thus the saving of them from death, has also been noted among *Stegodyphus saraindrum,* a communal spider. In this case laboratory researchers fed one spider mother-to-be with radioisotopes after she had completed her egg case. Later they discovered that the small spiderlings of another mother, unmarked by isotopes, were receiving the marked mother's food, which she was feeding them by regurgitation. To follow up this unexpected phenomenon, the entomologist E. Kullman switched egg cases on another commune member and discovered that this mother spider took very good care of the changeling spiderlings as they hatched. It seems that spiders—and other animals as well—evolve social mechanisms that enable them to suppress what appears to be an instinctive hostility between adult females. This occurs as the advantages of group living begin to outweigh the dangers.

Some primates, carnivores, and ungulates will suckle any infant of their species, even when their own offspring are still living. Observers in the mountains of India have noted that langurs will feed the offspring of any member of their group. And the mammalogist Ranka Sekulic observed this same kind of behavior among the almost extinct roan antelope in Kenya's Shimba Hills National Reserve. Antelope, like sable and giraffes, belong to "hider-type" ungulates. The mothers-to-be isolate themselves when they go into labor and then hide their new babies among the high grasses for the first weeks of life. The babies lie and wait for their mother's visits to feed them, early in the morning and again before her own afternoon feed. Sekulic noticed two mother antelopes that gave birth at about the same time, producing one male and one female calf. Soon the calves were suckling from either mother. And when the male calf succumbed to a leopard, the remaining female had two willing food sources, a real mother and a wet-nurse. Elsewhere in Africa, observers have reported female roan antelope nursing two calves at a time. These calves may have been twins, but more likely were an orphan and a mothered calf, both being taken care of by the same mother.

Because the willingness of a mother to suckle and guard an orphan has been observed in primates where family connections are close, and among herd animals that live in small, related groups, biologists speculate that the mothers that nurse orphans are closely related to them and are bringing up their own nieces or half-sisters, offspring in whose genetic success they have a vested interest.

However, entomologists have observed large-scale adoptions among neighboring but unrelated laboratory colonies of *Leptothorax curvispinosus.* The larger cluster of ants invaded the smaller group, killed the adults, and carried the immatures back into their nests, where they raised them with the pupae of their own young. This is understandable in the context of needing larger populations to meet an enemy, but it is not consistent with the theory that the main purpose of reproduction and care of the young is the perpetuation of the parental DNA, since the genetic makeup of the adoptees in these ants is different from that of their stepparents.

Communal Child Care

Females that are guarding young apparently sense that it is more economical to pool their energies. Birds like the groove billed anis lay their eggs in communal nests, as much for defense as for the convenience of sharing the feeding. Occasionally a pair of mother coatis, small Central American carnivores, sleep together on the forest floor. They do not share food or hunt cooperatively, but they forage as a team and sometimes care for their young in a communal nest.

Mother giraffes take turns watching a creche of their young in the shade while the others go off to find the leafy trees they need to forage from. During the daytime, a pair of giraffes stand guard to make sure that no lions or other predators invade. When the mother in charge senses danger, she moves to her own calf and leads it to safety. The other calves in the creche follow, although the mother only herds her own. The system she has become a part of enables all the calves in a creche to be well looked after by one vigilant sitter. Those in the creche stay protected from the heat of the sun as well as from predators, providing a long childhood for these unusual ungulates to develop the physiological advantages mandated by their long necks. Communal child care in this situation permits the lactating mothers within a group to share the responsibilities of the job, releasing the others to hunt or forage and in general do other chores which every mother must perform.

And among insects there are many nonsocial bees that live alone rather than in hives where specialized castes have formed, yet share maternal cares with other nonsocial bees. Each female provides for herself in most areas of life but builds her nest chamber in what would seem to resemble an apartment complex. Together with her counterpart loners, she shares a foyer where all can gather to fight off an intruder. Nonsocial spiders also pool their efforts when it comes to child care, and in the species *Dolomedes*

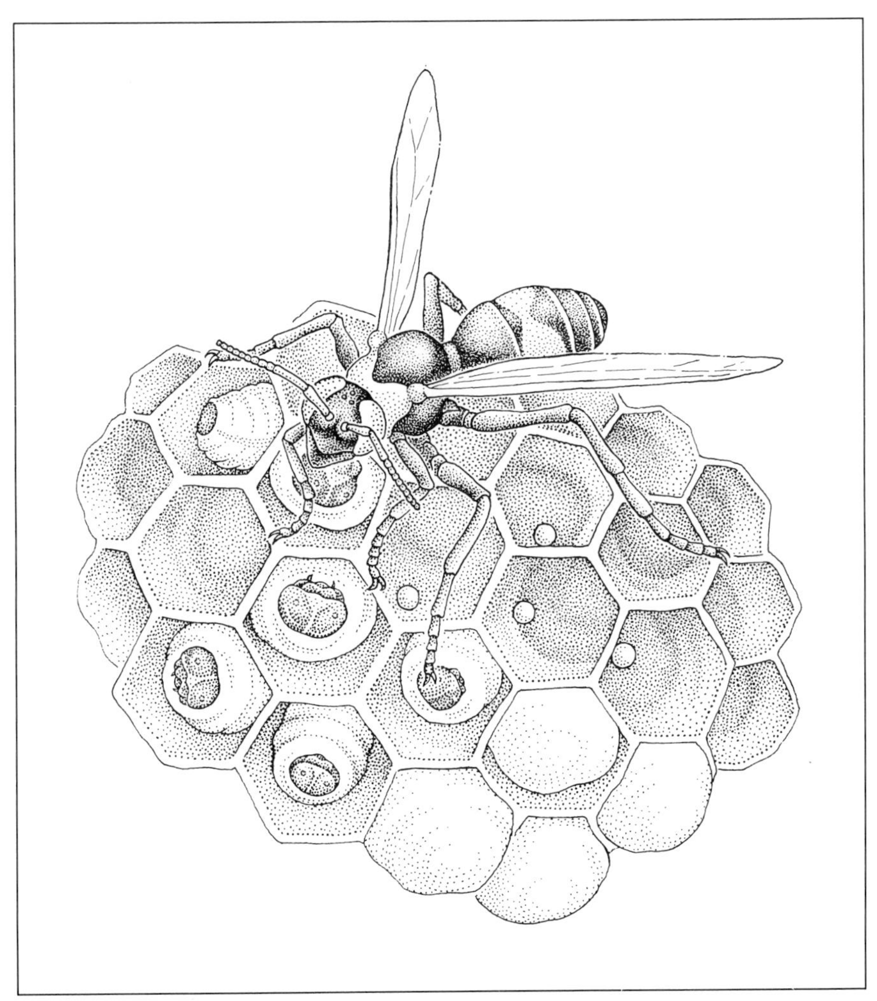

The adult female social wasp feeds a creche of growing larvae.

mirabilis they spin a nurse web where individual mothers lay their eggs and later raise their spiderlings.

Among the social wasps, where child care is communal, the adult wasps chew insects and then feed the masticated food to the growing larvae. The larvae are the only wasps able to transform protein into carbohydrates, and after they do, the young in turn feed their elders in a process called *trophallaxis,* by which the young secrete a sugary liquid from their saliva glands directly to the worker females.

Most social insects recognize their own, and each hive acts as an independent organism. Occasionally, however, some insect species, such as stingless bees *(Melipona),* meet in super colonies where several queens cooperate in subordonning literally millions of workers to raise an enormous number of viable offspring. In a similar manner, ant colonies thrive in especially arid areas where several cooperating queens divvy up the limited resources.

Communal care of the young is frequent among large African mammals. Prides of lions seem to orchestrate births in synchrony so that the three to twelve females among them can pool their efforts. According to Brian Bertram, who studied lions intensively in the Serengeti, the prides are stable; the females that control them never allow a strange female to join. Yet the pride seldom dies out. Birth to a pride member does not guarantee a place; at about three years of age some females become permanent members of the group while the others are expelled and become loners (since no other pride will accept them). These lone, isolated females bear smaller numbers of cubs throughout their lifetimes, most of which do not survive. Female lions within prides fare better. The pride provides a strong support system which enables the young to grow up protected. The pride females seem to come into estrus at the same time and bear their offspring together. They suckle one another's cubs, provide what seems like a mutual defense pact, and later bring back freshly killed prey as the cubs grow older.

Small animals like ground squirrels burrow holes in the earth, to insulate themselves from the cold in California's high Sierra Nevada and in Alaska's Yukon Territory. The Arctic ground squirrels enjoy a brief summer during which the females clump their litters, sisters or mothers and daughters joining forces. With their litters together, they have some insurance. In the event that one of the mothers dies, the remaining mother protects the orphans from the elements and from predatory males of their species, including fathers, which have a tendency to kill their young. The strategy of infant clumping is also found among wolf spiders that clump their egg cases, apparently for protection.

In the United States the wild bison female delivers her young in isolation from the herd in the same way as African ungulates. But once she is established as a new mother with a youngster at her side, she behaves very much like the African buffalo. The other females rally around her, encircle and defend her from meat-eating predators. And until the youngster is old enough to stand up for itself, the adults huddle close and provide it with warmth and protection.

It would seem that female animals are especially supportive of each other in social groups. Indeed, the relations between females in most species are more complex and enduring than those between males and females, even within species that live monogamously. In the few species

Female African buffaloes form a wall to protect the young.

that have been studied, within families the degree of cooperation between individuals seems to depend on the degree of kinship between them as well as the possibility of reciprocity at some later time. But the thin line between cooperation and competition is often hard to see. Among females of the species, the apparent selflessness that even cooperating sisters manifest may not be what it first seems.

thirteen

Cooperation of a Different Sort

Both in nature and in the laboratory, zoologists have noted in an assortment of female animals—some living with and some without any males of their species—relationships with other females that are clearly sexual. Although these behaviors differ and can be explained, in part, by peculiarities of the species, there is a connecting thread between the behaviors that tease the still unanswered questions about all higher forms of life. What is sex, after all? Why do animals need it and how did it ever evolve?

David Crews's exploration of the world of the parthenogenetic whiptail lizards *(Cnemidophorus)*, in the southwestern United States, reveals a pattern of male role-playing by one of two paired females. Crews notes a similarity between the behavior of females in all-female populations and that of males in occasional heterosexual pairs. A lone whiptail parthenogen will lay eggs that go on to develop. But two females together will produce more clutches more frequently than will an isolated female.

This species was probably heterosexual in the evolutionary past, and the male role-playing may be simply a nonfunctional vestige of the time these lizards were sexual. Or, as seems more likely because of the greater reproductive success of the paired parthenogens, heterosexual role-playing may compensate for the loss of the male-triggered stimuli that influence ovarian function.

Whatever the history, captive female lizards in cages fall into ovulatory syncopation. When one female is in the middle of her ovulatory cycle, her partner is just at the beginning. The nonovulating female takes over the role of the male. She becomes aggressive and lunges at her passive cagemate, gripping her partner's leg in her jaws. She then mounts her cagemate's back and strokes her neck with her jaw. Then she twists her own tail underneath her partner's in a position which is, in fact, the copulating position of sexual *Cnemidophorus* lizards. At this time the "passive" partner ovulates. As the month proceeds, the ovaries of the facultory lizard, that is, the female playing the role of male, begin to grow and the pair switch roles.

What this may indicate in terms of evolution is unclear. Crews has

Female lizards in mating position prior to parthenogenetic reproduction.

found, however, that females of parthenogenetic lizards produce more female hormones when approached by a member of their species that goes through the motions of being a male. The male-like behavior of one lizard stimulates the other to produce more of whatever is needed to manufacture eggs. However this behavior evolved, these lizards can be called truly cooperative. By mimicking the behaviors of their putative male counterparts, they help each other produce clones, all female of course, but quite capable of continuing the species.

Parthenogens are not the only kinds of female animals that play the part of males in their reproductive activity. Since 1972 a research team directed by George Hunt has been tracing homosexual activites among the western gulls *(Larus occidentalis)* on Santa Barbara Island off the coast of California. Although most gulls form heterosexual monogamous pairs and lay two or three eggs in a clutch, the team noticed that up to fourteen percent of the clutches contained between four and six eggs, most of which never hatched.

The unhatched eggs turned out to belong to paired gulls, of whom both were female. Like the heterosexual couples, the female pairs (identified and tagged by the researchers) remained together for consecutive seasons. They incubated their eggs as if they would hatch, and occasionally one would. (This points to a certain amount of promiscuity on the part of the male members of supposedly monogamous couples elsewhere on the island.) Members of the homosexual pairs exhibited most of the courtship and territorial habits of heterosexual gulls, including the occasional courtship feeding of regurgitated food, and even intermittent attempts at mounting and copulation.

The close intervals between egg-laying as well as the larger number of eggs in these pairs indicate that both females are probably contributing to the joint clutch. But the large-scale feeding of the female by a male is absent within the homosexual couples, which might explain why their eggs are on the whole smaller than in heterosexual nests.

Female–female pairs have since been discovered among ring-billed gulls *(Larus delawarensis)* and California gulls, though not as frequently as among the species in the Channel Islands. The population on Santa Barbara Island seems to be disproportionately female, and it may be that homosexual pairing is a strategy that allows those extra females to get fertilized by promiscuous males and then to incubate their young as a mated couple. If this is the case, the curious pattern on the island may find these birds in an evolutionary stage halfway toward polygamy. But of course this does not explain why there is a surplus of females in the first place.

In other species, females play the male role during a special part of their reproductive cycle. Among mountain sheep in the Canadian Rockies, which have been meticulously described by the ethologist Valerius Geist, young females occasionally mount each other, going through all the motions of copulation. When they are mature and in estrus, ewes become aggressive toward rams, butting and nuzzling them as they take the initiative in the courting process. It may be that the aggression they demonstrated earlier with younger ewes was a way of practicing the dominant role they play in heterosexual encounters. Likewise, female kob in Uganda move into the lek arenas, the mating territories of males, where they not

A pair of homosexual female gulls, captured and banded, whose reproductive behavior has been monitored.

only mount one another but exhibit the peculiar nasal contortions and leg-raising postures *flehmen* and *laufschlag.* This behavior was once considered exclusive to males just prior to copulation, be they kob, deer, or Australian wombats.

Another demonstration of female–female associations that became polarized into role-playing of male and female has been observed among orange cichlids *(Etroplus maculatus)* of the family Chromidae, freshwater fish native to India and Ceylon. The ethologist George Barlow notes that while he has not observed female–female cooperation in nature, he did contrive an artificial situation that stimulated females to act like males. In tanks from which he removed all males, the female cichlids formed pairs, each member of which spawned and helped care for eggs just like a heterosexual pair, except that these unfertilized eggs never hatched. However, when Barlow replaced the opaque separation between their tanks with clear glass, enabling the females to see the males across the water, the newly formed pairs dissolved. At the least the experiment proves that these cichlids rely on visual clues for mates, and that given an option, they prefer heterosexual pairings. But Barlow may have touched on an evolutionary principle akin to the behavior of the parthenogenetic lizards. Given a gradual disappearance of males over a long period of time, some of these female fish might also become parthenogens, caring for their eggs and fry as parents, mother and father. Conceding that the laboratory situation is contrived and that these fish were peculiarly stressed, one may say that in an evolutionary sense, sexual behavior persists even when sexual distinctions do not.

Mutual stimulation by female lizards or mountain sheep prior to the arrival of males is understandable as a way females help each other prepare for motherhood. Less understandable in a reproductive sense is the behavior of female pygmy chimpanzees. This small chimpanzee behaves in ways more reminiscent of *Homo sapiens* than those of ordinary chimpanzees, gorillas, or orangutans. They have thus provoked a great deal of fanciful hypothesizing from those zoologists and behaviorists, who are always alert for some manifestation of a behavioral "missing link."

Pygmy chimps have a neotenic appearance; that is, they retain into maturity the head shape and expressions of younger animals, just as humans do. They also copulate face to face. In addition, females of this species engage in frequent mutual masturbation. With one hand, each animal holds her genitals open so that her clitorus extends to touch her partner's. With the other, each holds her partner about the hips as both swing their hips from side to side. A session of this behavior, labeled "GG" (for genital–genital by the primatologist Sueihisa Kuroda, who first described it) lasts about twenty seconds, slightly longer than heterosexual copulation in this species. Kuroda believes that female pygmy chimpanzees perform this act as a way of relieving tensions. He notes that it frequently occurs after a chimp that has food has given some to a food beggar. At the least, he suggests it proves that there is friendly behavior between females, cooperative behavior in a manner that has not been observed in other nonhuman animals. By using each other as an outlet for whatever tensions they may be experiencing, these primate females are helping each other in a way that is not immediately connected with either reproduction or the care of offspring. In this instance, we glimpse a sexual

Female pygmy chimpanzees in Zaire engaged in "genital-genital" stimulation prior to heterosexual copulation.

dimension of animal life beyond the overriding concern with genetic immortality.

A closer look at female homosexuality among primates is provided by Linda Wolfe, who spent a decade examining the behavior of a captive colony of Japanese macaques near Laredo in West Texas. The Laredo group, referred to as Arashiyama West, were originally part of a genetically related group (referred to as Arashiyama B) that remained intact in their natural habitat in Japan. When the groups diverged, their populations changed, and so did their behavior. The Japanese colony, which was provisioned in the wild, grew in size; but the transplanted monkeys dwindled in number as some animals seemed to die prematurely and were not replaced by either migratory males (for there were obviously none in Texas) or newborns.

Observers had earlier noted a small degree of sexual activity between adult females, even in Japan, but the females in their natural habitat also had heterosexual relationships with migrant males who only appeared during the breeding season. In Texas, Wolfe noted an increase of sexual activity between adult females, an increase numerically large enough to make it appear as if the relocated females supplemented heterosexual relationships with homosexual ones.

Wolfe observed consortships, exclusive relationships between individuals, with prolonged sexual interactions. She noted that the female–female mountings were confined to the breeding season, coincident with heterosexual copulations, and that the female consorts followed each mounting with mutual grooming, huddling, and foraging in the same way that heterosexual macaque couples consorted. Moreover, she noted that the females did not form sexual relationships with close female relatives, just as the heterosexual couples also avoided such close alliances. And finally, she observed the females rubbing and pressing against each other and occasionally making eye contact, grimacing, and clutching—behavior that some primatologists have argued indicates the existence of orgasm in these nonhuman primates. Wolfe argues that female homosexual interactions increased in the Texas macaque colony because there was a dearth of novel males in the community. She calls the result "novelty deprivation homosexuality." Yet even in Japan these female macaques indulge in homosexual mountings and frequent masturbation, indications that sexual stimulation and procreation are separate behaviors in this species, as well as in the pygmy chimp.

fourteen

The Other Side of the Coin

LIFE IN A GROUP can enhance the chance of reproductive success for an elephant or a honeybee, providing help in the constant search for food and protection. Beyond that, group life offers the intangible advantages of companionship, and perhaps an outlet for stress through the reassuring presence of a familiar individual. These advantages for the female have been noted by generations of zoologists. However, many of the subtle tensions and stresses that result from living within a social structure had, until recently, been overlooked. Within groups of females in a large number of species, individuals do not have an equal chance at survival. For occupying the lowest rank in a hierarchy can prove disastrous when times are poor, the food supplies sparse, and mates, for some reason, unavailable. If there is not enough to go around, females with low status suffer disproportionately.

Dominance hierarchies among females are more distinct among polygynous species, in which just one or a few males share the company of several females. Within large groups, individuals of both sexes often live in hierarchies. Although the rankings can be subtle, each animal, male and female, appears to know its place and is reminded of it every time it makes a move, whether seeking a spot in the shade, a partner in copulation, or simply the right-of-way on a narrow path.

As a young student of animal behavior, Solly Zuckerman, analyzed the baboon colony at London's Regent's Park Zoo in the 1930s. Looking exclusively at the behavior of the males, he concluded that baboon societies are organized into a pyramid with the alpha male at the top, surrounded by a cluster of subservient females. This description was an important landmark in the study of animal behavior, because it spotlighted the key role that rank plays in animal societies. But it turned out to misrepresent the relations among male baboons when not in captivity; and by ignoring the behavior of females altogether, it greatly distorted the role that female competition plays in baboon society. There are many subspecies of baboons living in Ethiopia, Kenya, and Tanzania that have been followed in a series of field studies for the past twenty years. Their patterns of behavior differ, yet among males there seems to be an alpha personality, a leader, at any particular time, and in all groups there are both male and female hierarchies.

Samuel Wasser's study of yellow baboons in Tanzania reveals the considerable power that high-ranking females have over the reproductive success of their subordinates. Zoologists interested in measuring fertility had at one time simply counted completed copulations alone, which were distributed fairly equally among females. They had ignored the number of births and the subsequent development of offspring. But every copulation does not result in conception with monkeys any more than with *Homo sapiens.* When higher-ranking female baboons, which inherit their rank from their mothers, gang up on lower-ranking ones, they usually attack as the victim's estrous swelling begins. They seem to be deliberately interrupting the reproductive cycles of the victims. For the attack disturbs ovulation, and any copulation that follows will likely not result in conception.

While observing a group of another kind of baboons, olive baboons near the Gilgil Rift in Kenya, the primatologists Barbara Smuts and Nancy Nicolson happened upon a curious palace revolution. The females had been fighting each other, subtly with threats and blatant face-to-face combat, when the observers had to leave. They returned some months later to find that the four females at the top of the hierarchy had fallen to the bottom. The researchers deduced that when a key, high-ranking female had accidentally drowned, her death triggered the dramatic changes in the status quo. More important, having observed the same four female baboons as leaders at the top of the heap, and later as victims at the bottom, these researchers could not distinguish any real behavioral differences between the individuals in their demeaned circumstance, save for the fact that, in their new roles, the now low-status animals were careful to avoid harassment. Baboon hierarchies, though rigid, do change. And the females play out their roles, apparently, because of where they are on the totem pole rather than because they have some inherited genetic talent for either leadership or victimhood.

Jeanne Altmann's study of mother–infant relations among a group of yellow baboons in Kenya revealed that maternal styles vary according to the mother's position. The highest-ranking mothers let go of their youngsters early, encouraging their independence, because they could apparently count on support for them from the lower-ranking females the youngsters would encounter. So the daughter of an alpha female baboon was able, when her mother died prematurely, to carry on by herself. But the lowest-ranking female clung to her child, protecting it because what she could expect were attacks and assaults on her offspring. Eventually, of course, the youngster had to face the world alone, but this mother post-

A mother hamadryas baboon carrying her youngster.

poned that day, protecting the youngster while it grew stronger and better equipped to defend itself from its low perch on the totem pole.

Each animal society is different. Among primate species such as India's Hanuman langurs, a female's rank changes with her age. As she matures and bears her first offspring, she reaches the top of the heap. But as she ages she loses status. Past childbearing, she lives on the lowest rung of her hierarchy and plays out a special, sometimes hazardous, role as sentry, aggressively defending the troop from dogs and interfering humans, as well as protecting infants from the assaults of other langurs.

Among the anthropoid apes, gorillas give priority of place to a new mother, who then slowly loses status as her infant matures. Later, if she becomes pregnant again, she will move back to the top, which among gorillas means into a resting place closest to the silverback male who is probably the father of the youngster.

Females of other species do not appear to care a whit for either age or reproductive stage but maintain allegiance to that female's rank at birth. Hyena mothers pass along their status directly with birth to their daughters, the adults reaffirming their rank with a complicated greeting ceremony whenever any members of the group meet. Daughters of dominant mother elephants also inherit their mother's position and, unless they are sickly, retain that position throughout their lives.

When animals are as long-lived as elephants, the dominance hierarchy extends over generations. Not only do the leaders apparently suppress ovulation in the adolescents, keeping them as long as possible in the role of nursemaids, but as they grow older, females with high rank may be able to control their own ovulations. The highest-ranking come into estrus ahead of the others, and so bear their young earlier in the season when food is most plentiful. Even when pregnant, these older elephants continue soliciting the sexual attention of males as a way, it has been suggested, of preventing those same males from impregnating the younger females, whose offspring would inevitably compete with their own.

Among rhesus macaques on Cayo Santiago Island, daughters inherit their mother's rank, standing above all females to which their mothers are dominant, and below those to which she is subordinate. A macaque daughter does not outrank her mother, but the youngest sexually mature daughter outranks her older sisters. Thus a female's rank declines slightly after her birth if her mother goes on to produce younger sisters that the mother then favors. On La Cueva Island, off the coast of Puerto Rico, the primatologist Z. C. Drickamer conducted a ten-year study which revealed that female rhesus macaques of higher rank not only begin to bear their babies earlier than lower-status females, but they continue to bear offspring at shorter intervals and thus produce more living offspring. These high-born offspring have a greater survival rate than the infants of lower-ranking monkeys.

Low-ranking female macaques suffer early and frequent harassment. Higher-ranking females seem deliberately to bite and shove young females unlucky enough to have low-ranking mothers. Yet they withhold the same kind of abuse from the young female's low-ranking brother, which, it would seem, ranks equally low. The focus of the offenders is clearly other females that could become competitors. They ignore the males, which in the natural course of events will leave the community. Studies of captive pigtail macaques, cousins of the rhesus, reveal that the lowest-ranking females require the most medical treatment from keepers, a result of bites inflicted on them by their higher-ranking cagemates. Moreover, the high-rankers produce many more infants than the scarred victims. When subjected to artificially induced stress, low-ranking macaques experienced a threefold increase in spontaneous abortions as compared with the high-ranking females among them.

Primates are not alone in harboring females that abuse one another. Small raccoon-like coatis also pal up, female with female, to subdue or simply get rid of other females they find obstructive. And studies of Norway rats reflect a similar hierarchical pattern. While higher-ranking females successfully wean all their young, low-ranking rats, easily visible by their scars, eat more than 60 percent of their offspring. When the mothers cannot find food, they cannibalize their infants, which would starve to death anyway.

Subordinate female mammals are clearly at a disadvantage. They must cope with a great deal of aggression in order simply to come into estrus, copulate, and bear an infant to term. Yet they are not without weapons to retaliate. Among those female mammals where daughters remain and sons migrate, the one ploy a low-ranking female elephant or baboon can use is to produce more sons than daughters, which is precisely what she does. Migrating sons apparently start out with no status encumbrances, but daughters remain "home" to reenact their mother's role. Zoologists have recorded that among elephants, baboons, and bonnet macaques—all species with strict matrilineal hierarchies—low-ranking mothers produce significantly more male than female babies, and high-ranking mothers produce more females. The mechanisms that affect the sex of offspring are unknown. Equally mystifying is the discovery that pregnant baboons that were harassed into miscarriage in situations where their fetuses could be examined turned out most often to miscarry female offspring. It is as if the attacking females sensed the presence of a potential rival, or else the harried mother reacted to stress by cutting her losses.

Hierarchies are not necessarily life-long, as we saw in the olive baboons. Elephant seals spend only a few weeks a year on land. The rest of the time, as far as we know, they swim about solitarily, the females carrying fertilized eggs with them which, months later, begin to develop, apparently triggered by the change of seasons. When the seals eventually arrive on the islands where they will, within days, give birth, they arrive as strangers to each other. But a hierarchy quickly forms among them, dependent, apparently, on age (which observers can determine by the number of bite scars on the female's neck). The oldest mothers have the highest rank and demonstrate it by attacking and killing newborn pups of their younger competitors. So it may be several years before a young female becomes a successful mother. In the meantime, she gains experience, learning to establish herself so that she can, in turn, raise pups of her own while killing the offspring of younger seals that have matured in the interim.

Because these temporary hierarchies are age-dependent rather than birth-dependent, sometimes a mother's most serious competitor is her daughter. Female deer cannot dive into the sea to escape the competition. What they can do in self-defense is see to it that if they bear sons, the sons receive more and better care than daughters. For instance, mothers of males devote more milk to them, as is shown by the relative size of male calves compared with females born at the same time. The male deer, like male baboons, leave their natal territory as they mature, and only the largest and strongest among them will return to impregnate a host of does. So the female will have a greater chance of becoming a grandparent many times if her son becomes the next season's accepted buck. In contrast, a daughter will become her mother's rival for the limited amount of territory available and can produce only a few offspring in comparison with a successful buck.

Dominance hierarchies within female populations had been observed among vertebrates for some years before entomologists recorded similar behavior in ants. In Massachussetts, the ants *Harpagoxenus americanus* live inside hollow acorns. The small nests include a single queen ant and ten workers that in turn enslave about two hundred ants from a kindred genus, *Leptothorax*. With extraordinary dexterity, researchers marked the

A dominant female slave-maker ant bites the leg of her subordinate sister.

worker ants with paint and monitored their behavior under the microscope. The results revealed a social hierarchy, including ritualized fighting among the workers, complete with antenna duels. The area of contention was competition for food, which the workers could obtain only from their slaves. The higher-ranking workers thus got the most food, and in turn produced the most eggs. These ants reproduce both sexually and parthenogenetically. The higher-ranking workers produce more males, which give these females a better chance of passing on their own genes into future generations of the colony. By removing the highest-ranking worker, experimenters were able to discover that the queen recognized the individual worker by rank and changed her relationship to each of them when the top-ranking one was removed. This remarkable view into the microscopic world of insects offers evidence that female animals tend to evolve similar strategies dependent on the social units in which they live, regardless of their phylogeny.

Subtle Aggressions

The most obvious way in which dominant females destroy their female competitors is through fights, especially those in which the winning team outnumbers the losers. Biting and clawing are not rare between females of many species. Baboons and macaques, lizards, and moorhens do not hesitate to attack their female competitors. Dian Fossey in Rwanda reported discovering a badly wounded young female gorilla at a time when there were no strange males around and when the only possible attacker was another, older female with two young daughters. Female redwinged blackbirds *(Agelaius phoeniceus)* in Duchess county, New York, are promiscuous breeders. During the breeding season these females turn aggressively on one another, not apparently to guard their territory but to win and maintain the allegiance of the siring male. Although they incubate their eggs on their own, they depend on the male to help with the offspring. In this species, where females outnumber males, which also happen to be providers, the females compete aggressively for male attention.

Most females, however, do not resort to the kind of fierce battles of hoof and antler that Darwin and his contemporaries observed (and exaggerated) in male animals. On the whole, competition between females tends to be subtle. The dominant female, be she baboon, rat, elephant, or woodpecker, exerts pressure on subordinate females, or perhaps just younger ones, that results in either sterility or simply failure to attain sexual maturity. This strategy is noticeable among certain birds, where females live in family groups dominated by territorial males. Most of the females eventually migrate, but often young females remain at the nest, bringing food to their younger siblings and generally helping out. During this time they are sexually mature but do not lay eggs. The presence of both mother and father apparently prevents ovulation. Some of these females in species like the Florida scrub jay *(Aphelocoma coerulescens)* remain at home until they are no longer young, and never breed at all.

Ornithologists have considered this situation among several kinds of birds and suggest that in each instance, the nonbreeding female somehow assesses the costs and benefits of going off on her own versus remaining behind to help with her siblings. In some instances, as with some American jays in the southwest, there is simply no territory left for nesting. Many females cannot find places to make their own nests, so they stay at home as though awaiting the unlikely vacating of the parental nest that would come with the death of the parent birds. Female jay helpers cooperate in bringing food to the hatchlings, but they have also been observed to destroy the eggs in their own group's nests or in nests in adjacent territories. Helpers may not be really helping when they sabotage the breeder—in frustration, perhaps, at not being able to start their own family.

Monogamous bird families are reminiscent of monogamous mammals that live together—mother, father, and children of assorted ages. These family groups are often mother-dominated. Within families of marmosets in the Amazon forests, or British foxes or African wild dogs, the mother's presence seems to inhibit the sexual maturation of the daughters. There is only one reproducing female, regardless of the age of the daughters. In captivity, researchers have proved that young marmosets only begin ovulating when they are removed from their mother's cage. Because of some mechanism that most likely responds to stress, the younger females do not mature completely until they are on their own. When marmosets live where there is enough room to emigrate, the females leave after a year or so to find a mate and begin rearing their own families. Where there is no place to go, the daughters can and do retaliate—like the young jays that destroy eggs in the breeder's nest. Their apparent clumsiness draws the attention of predators to the helpless young. Other helpers are heavy-handed and harm the young they are supposed to aid, which seems to be the price a mother pays for dominating reproduction in her territory.

This kind of control, where mothers apparently exert pressure to arrest their daughters' sexual maturity, occurs occasionally among mammals and birds, but routinely among social insects. The queens in a bee hive, for instance, produce large amounts of a pheromone that renders all of the other developing females sterile. If the queen dies, the workers sense the absence of her pheromone and quickly begin feeding royal jelly to some of the larvae in order to rear a new queen. But the control is mutual. According to recent studies, the pheromones sent out by queen ants have a specific intensity. Groups supporting only one queen band together and kill a second queen that has a different smell. Groups with several queens can only tolerate a certain quantity of this control chemical, and when that level is reached they react by killing any other queens that try to operate within the nest.

Suppression of maturation among social insects is one way that some females control other females to get their help and forestall their competition. The process is especially clear among vespine wasps. The queen is not only singularly larger than the sterile workers, but she is thoroughly developed. Not so her workers, whose vaginas are too small for copulation and whose ovaries are incomplete.

Beyond the world of social insects, observers have found that African naked mole-rats *(Hetercrophalus)* also live in large colonies of more than forty members with a single breeding female. Evolutionary biologists believe that this is the first vertebrate group that can be identified as *eusocial,* a society in which the different functions are carried out by classes that differ physically from one another. This social structure had previously been thought to be confined to the world of insects, particularly those with haploid-diploid genetics. Fascinating to entomologists as well as mammologists, mole-rat society resembles a termite hill more than any vertebrate family. Like termites, the mole-rats live in a very tight society in which only one female in the whole group reproduces. The nonfertile females do not develop prominent teats or a perforate vagina but have the look of perpetual subadolescents. The rest of the colony is divided into castes determined by their jobs. Like the termites, the workers are smaller than the mother and fathers. They forage for food and keep the tunnels of their burrow in good repair. The nonworking males spend a lot of time sleeping, but even asleep they are still working because their body heat is what keeps the newborn members of the colony warm. And when danger threatens, males join the female workers in carefully carrying the young to safer ground.

Delving into the physiology of these mammals, most of whom never mature sexually, the zoologist Jenny Jarvis in South Africa discovered that the large matriarch (or alpha female) transmits a substance that prevents the rest of the females in the colony from maturing. She patrols the complicated network of tunnels they live in, obliging other females to lower

The alpha female naked mole-rat (below) and a subservient female inside their burrow.

themselves to let her pass and to press their noses against her own. In this great labyrinth they share a communal toilet where, it is possible, the mother's urine contains a pheromone that separates the colony into castes. From the toilet's strategic location she is able to affect all members of the colony.

Simply harassing a subordinate female can make her forego estrus altogether, suffer a miscarriage, or in the case of the gelada baboons observed in Ethiopia, endure months of sterility. As if making a double-check to ensure their reproductive dominance, many female mammals actually interfere with their subordinate's efforts at copulation. Among baboons, in what appear as carefully planned actions, coalitions of higher-ranking females actively interrupt when a male tries to mount a subordinate female. Zoologists have seen similar behavior among wolves at the Brookfield Zoo in Chicago, where the dominant bitch assaulted the other two females physically as well as by threat whenever they solicited a male or were receptive to one. The same kind of female interference in copulation has been observed among three-spined stickleback in both streams and aquaria. Dominant females apply pressure to the spot on their subordinate peers where a male fish would normally stimulate the female to spawn. In a heterosexual encounter the male would immediately fertilize the eggs. But in this instance the subordinate female releases her eggs and, since there are no sperm to fertilize them, they are wasted. This behavior, called *female quivering,* reduces egg competition for the dominant female, who has arranged it that her fry outnumber her rivals in the pond.

In most egg-laying species, females that have not successfully prevented insemination can wait for the eggs to be laid, then deliberately destroy them. Like lions, female ostriches share their males, but where female lions seem to be democratic, ostrich society divides into one major hen and one to six minor hens and a single male. The female ostriches collaborate in the savannah, digging one shallow dish into the ground which is all they will have for a nest. Between the lot, the females will lay about forty eggs, but there is only room under the incubating bottom of a grown female for about half of them. This means that half the eggs are lost almost as soon as they are dropped. The major hen usually lays just a few, her co-tenants many more. Yet the dominant hen is able to recognize which eggs are her own and she rolls some of the others out of the way, ensuring that they will not develop. How she recognizes her own eggs, whether by smell or size or even surface scraping, is unknown. But experiments marking each egg according to mother have demonstrated that somehow she does make the distinction.

Geographically distant from East Africa, the South American anis *(Crotophaga sulcirostrus)* also share nests with other females, and also do away with "extra" eggs. Before she lays her own in the nest, each female rolls out a few of the eggs already there, then sits a spell at incubation. As each female tosses out eggs before she has laid her own, she has no need to identify the others but instead tries to be sure that her own are the last to be laid. The last to lay is clearly the winner, and that place usually goes to the strongest, oldest bird. Among other birds that compete for male attention, like the buttontail quail, females blatantly seek out and destroy other females' eggs—a technique called egg-rolling.

Two options that a female has when it comes to increasing the proportion of offspring that will survive into the next generation are to breed as rapidly as possible in order to out-produce competitors, or see to it that her competitors' offspring do not survive. This last strategy offers several opportunities. If she does not successfully prevent her rivals' maturation, she can attempt to stop ovulation, and if not ovulation, then insemination. Failing all of that, she can resort to deliberate infanticide. This is a very effective and common way of providing space for her own young in the competitive world they will enter. Most of the infanticides that have been observed, especially among mammals, were inflicted by fathers on offspring they sensed were not their own. That kind of killing is distinctly at odds with the mother's interests. When a female kills an infant that shares the same father as her own offspring, as happens with polygynous birds, the female is, in fact, returning the favor, so to speak, to males that in other situations destroy their adversary's offspring. She is removing the competition of an infant with which she shares no genetic material so that her own offspring can benefit from whatever the environment has to offer. The infanticidal female sees to it that what resources exist—food, protection, or sexual partners—go to her own offspring no matter which animal the father happens to be.

Infanticide does not occur among solitary animals. It is only useful when animals are living together and sharing what sometimes turn out to be limited resources. The same females that have depended on each other for assistance, sharing a nest or an earth burrow, want the best place in that nest for their own eggs, or the best part of the burrow for their own pups. Belding's ground-squirrel mothers will quickly kill any babies they find in an unguarded burrow, and female lions and jackals will seek out and kill cubs that for any reason are not being carefully defended by their mothers. These killings have been rationalized as acts of mercy, the killers sensing that the delinquent mother is dead and that the offspring are doomed. Or they can be looked at as a way in which the mammal mothers cut down the numbers with which their own offspring will have to compete. Whatever the motivation, infanticide by females is the leading cause of mortality among young ground squirrels and is a high contender for infant mortality among wolves, chickens, and geese.

Occasionally, the mother, having produced a previous youngster, somehow encourages the older sibling to kill the younger one. Eagles, we know, sometimes let the first-hatched dine on the back-up chick. Among jackals, the female helper has been observed killing younger siblings when food has been short. Among wild dogs, a second female is often permitted to become pregnant and bear pups, yet those pups are killed by the dominant female.

All infant deaths are not infanticide. Some look like accidents, and may well be, yet they occur often and seem to fall into a pattern. Among primates, a good number of youngsters seem to die by mistake. Infant vervet monkeys may become victims in a tug-of-war between mother and "aunt." Other monkey infants have been used as buffers in fights. The presence of an infant, especially one whose color and markings still flag it as an "untouchable," can forestall the advances of an angry male. But often enough the buffering tactic does not work, and the infant is flung aside and wounded or killed in the ensuing combat. And occasionally the infant is simply "aunted to death." That is, the curious adolescent female "borrows" the baby, then loses interest but fails to return it to its mother. If the infant does not get back to her in time, as with small marmosets that need milk as frequently as every four hours, it will die of dehydration. Occasionally female baboons will deliberately kidnap the offspring of a very low-ranking mother, tossing it back and forth, keeping it from the mother until it, too, has died from being mauled and starved. When considered as isolated instances, these kinds of infant deaths seem accidental; but there are so many and the victims are so often the offspring of young or low-ranking females that it is hard to interpret these deaths as anything less than a devious form of infanticide.

Some female animals protect their young even if that obliges them to become aggressive. These same females will cooperate with each other as long as that apparently suits their long-range goals for healthy heirs. But they will change their behavior if their offspring seem endangered.

At times, a female animal will appear to behave destructively for no apparent reason. Observers of primates describe harassing behavior among the gelada baboons as simply "spiteful," acts perpetrated by one individual on another in which the attacker harms herself as she harms another. This kind of behavior has also been observed in three-spined stickleback and among birds like the greater prairie chickens, a lekking species that inhabit a great deal of the west central United States and Canada. In early April, the female chickens fly in small groups to the males' breeding ground. The females arrange themselves in a hierarchy as the birds return to the lek on three or four succeeding days before they copulate with the chosen male. The dominant female in each group ag-

Adult female gelada baboons in a tug-of-war with an infant.

gressively prevents the others from copulating until she has made her own choice. Observers recorded one instance in which a single dominant chicken interfered for four days with the successful copulation of her subordinates, an especially harmful behavior for a species in which the viability of a chick depends on how early in the season it is born.

Spiteful behavior has been explained as a social act that detracts from the perpetrator's benefit as well as the victim's. Ultimately, however, the perpetrator seems to anticipate a long-term advantage for which the price is a short period of discomfort. Of course animals cannot be expected to reason this way, but they appear to have learned somehow that spiteful behavior is both emotionally satisfying and in the long run beneficial. It is possible that "spiteful" actions may be simply another way of eliminating competition before it appears and is, in fact, of considerable advantage to the so-called spiteful female. Or they may be an independent kind of behavior in which females mistakenly underestimate the ability of the habitat to support a large population. They thus cut down the potential size of their own species, even if they may be at a disadvantage in the face of competing species should their numbers dwindle. Whether the aggressive acts that occur among female animals are really spiteful or are simply another way of controlling the size of the competition, they are a blatant demonstration of a lack of sororial support, an ascendence of competition over cooperation.

Females seem prepared to help other females of their species if they have some familial connection or in some other way may benefit from their gestures. All female behavior, like the actions of males of the species, is self-centered. Individuals use and abuse each other, depending on which kind of behavior will help them survive.

Epilogue

LONG BEFORE ethologists ventured to investigate animals in their own habitat, human beings had removed some species from the wild. Among them were honeybees, which we have sheltered, if not domesticated, for the past five thousand years.

The Egyptians understood the beehive as a social system akin to their own, with the largest bee, the leader of the hive, a super-pharaoh. Centuries afterward in Italy, the Roman Seneca described his ruler, Nero, as an "Emperor bee," an absolute ruler. Christianity spurred bee-keeping in Europe, largely out of a need for beeswax candles, and clerics then used the hive as a metaphor for the Church, especially the celibate monasteries where everyone had a place. It was not until Elizabeth I's reign in the sixteenth century that a contemporary Spaniard recognized that the king bee in a hive was really a queen. And so, at last, the lead bee was acknowledged as a female of her species.

Yet mankind went on insinuating its prevailing political views into the hive. In Stuart England, the queen bee was described in 1609 as an absolute monarch; then fifty years later, after Cromwell had dispensed with the idea of the divine right of monarchs, an Englishman wrote of "The Reformed Commonwealth of Bees." Language for the hive has continued to move with the political pendulum; after 1917 the Russians no longer referred to the queen bee as "tsaritsa," but called her simply "matka," mother, instead.

As with bees, so it has been with much of the animal kingdom. Naturalists in the past often read their visions of human society into the lives of animals. Darwin's perception of female choice as crucial to sexual selection met with skepticism in 1871; and even now, in spite of increasingly solid evidence of the female's role in determining which male shall father her offspring, there is still some resistance to recognizing the powerful role of the female in the evolutionary process.

While we share the interests of our Victorian predecessors in the evolution of animal behavior, we approach the animal world with new tools, conceptual as well as technological. We realize that understanding communities of animals is different from understanding the interactions of chromosomes. Animals are born, mature, and die; along the way, they incorporate emotional and psychological experiences into their patterns of behavior. But we are uncertain as to how flexible these patterns are, how much is instinctive and how much learned, within the different species.

We want to know more about these behaviors in terms of nonhuman animals, of course. But we are undeniably curious about how they pertain to ourselves. We are part of the continuum of animal life and at the same time different from even our closest hominid cousins by an order of magnitude that is hard to define but manifestly evident in the civilizations we have constructed. We cannot assign consciousness as we know it to other species, but we cannot ignore the fact that other animals share with us what appear to be strategies aimed at survival and reproduction.

Which leads to a subject that has hovered, unmentioned, over this book: What does the behavior of female animals tell us about human behavior? It is almost impossible to read about the cooperative child care among giraffes, the incestuous relations of woodpeckers, or the sexual abandon of pygmy chimpanzees without wondering if there is some message here for ourselves. How easy it might be to point to examples of infanticide, or cannibalism, or incest and decree the behaviors acceptable because they are "natural."

Polemicists can find analogies and, if they want, justification in the behavior of some female of another species for almost any type of human behavior. Even among our close relatives the primates, we can identify behaviors to justify, if we would, male domination and harems (as in langurs) or devoted monogamy (as in gibbons). Or we might celebrate the fact that females help other females and appear to enjoy family ties among birds like the white-fronted bee-eaters and such mammals as hyenas; or we might deplore the fact that other females, like England's red

foxes, appear to suppress maturation in their daughters to prevent them from having families. We could use these animal examples to suggest direction for human behavior, but we would be wrong.

What these studies do offer us is insight into the behaviors and the chemistries of other species so that we can learn about the forces that shaped our own development. We can learn what our ancestors might have been like by examining surviving species on different branches of the evolutionary tree, always bearing in mind that they are not mirrors of ourselves. We can conjecture about the development of some of our behavioral traits, such as territoriality or constant sexual availability; but there is no directive in the lives of other animals as to how we *should* behave.

When Darwin tackled the heroic task of analyzing all of animal behavior, he, like his contemporaries, unwittingly projected some of his own views of human females onto female animals. Today, we are in jeopardy of doing the reverse—appropriating patterns that come from the study of animals as imperatives for human actions. No matter how familiar we find the aunting of wild dogs or the spitefulness of baboons, we should not forget that the individuals involved have very few options as to what they do. Although our behavior is of course influenced by genetic constraints, it is far from being wholly determined by them.

It is valuable periodically to stand back, as I have sought to do, and compare the behaviors of very different species. The appearance of analogous behavior in animals as removed from each other as naked mole-rats and termites suggests that convergent evolution has occurred not only in different geographical areas but among different orders, not only in morphology but in social systems. The new studies of female behavior have done more than fill a gap in the puzzle of evolution. They have altered our understanding of how males live as well, and of how entire social groups, including the nonreproductive members, fit into the evolutionary picture.

George John Romanes assured his readers that "every Jack has a Jill." His mentor, Darwin, had assumed that most adult animals reproduce. Yet we know now that in many species, from ants to shorebirds, there are large numbers of nonbreeding adults that nonetheless play an important role in the survival of their kin. It is apparent today that evolution is cooperative—within the sexes and between the sexes in many species.

No one today would seriously suggest that female animals are mere egg repositories waiting for something to happen. Just as we are coming increasingly to appreciate the diversity of female roles in human society, so we are coming to understand the variety in the behavior of female animals—and to recognize females as, at the very least, co-equal players in the evolutionary game.

glossary / bibliography / photograph sources / index

Glossary

Adaptation: Any change in the structure, physiology, or behavior of an organism that allows that organism to function better within its environment. For a species, any change (usually genetic, selected for by natural selection) that allows the species as a whole to cope with an environment.

Alloparent: Any individual of the same species which assists parents in the care of their young.

Altricial young: Animals that are helpless at birth or hatching and therefore require parental care and feeding. (Contrast with *Precocial young.*)

Altruism: Any behavior that benefits another organism at a cost to the actor, usually where cost is defined in terms of reproductive success.

Bonding: A close relationship formed between members of the same or different species.

Brood: A group of young animals of any species that is being cared for by adults.

Brood parasitism: A situation found in some birds and insects in which the eggs of one species are put into a nest of another species, with the result that the host rears the young of the parasitic species as though they were its own.

Cannibalism: The act of eating the body of another member of one's own species.

Chromosome: A complex structure found inside the nucleus of a cell, carrying the genes—the basic units of the cell.

Clone: A group of genetically identical organisms derived from a single cell, or a population of individuals derived asexually from a single parent.

Competition: The demand by two or more organisms or species for a resource that is in limited supply, such as food, resting sites, or sexual partners.

Convergent evolution: Processes by which dissimilar organisms become similar with respect to a particular structure or behavior; for example, the appearance of fins in sharks and whales due to similar selection pressures.

Dimorphic: Having two different physical shapes, not connected by intermediates, in the sexes of one species. (Contrast with *Monomorphic.*)

Diploid: Having a double set of genes and chromosomes, one set from each parent.

Displacement activities: The performance of behavioral acts, often in situations of frustration or indecision, not directly relevant to the immediate activity.

DNA (deoxyribonucleic acid): The basic hereditary material of all kinds of organisms—in animals, located within the chromosomes in every cell—that codes for the replication of the entire organism.

Dominance (behavioral): A relationship between two individuals in which the subordinate individual withdraws or behaves in a submissive manner toward the dominant individual.

Dominance hierarchy: A "pecking order" of behavioral interactions in which every individual in an established troop, flock, or group is dominant to those lower in the order and submissive to those higher. The hierarchy is initiated and enforced by hostile behavior.

Entomology: The scientific study of insects.

Estrus: The period of heat or maximum sexual receptivity in some female mammals, usually at the time of ovulation.

Ethology: The study of animal behavior in natural environments or as free-ranging organisms in the ethologist's environment.

Eusocial animals: Groups of organisms displaying all of the following traits: cooperation in rearing the young; reproductive division of labor, with some members of the group usually sterile; an overlap of at least two generations of life stages contributing to colony labor.

Evolution: The long-term change and speciation of biological entities brought about by the differential survival and reproduction of individuals.

Extinct: No longer present in the population of living animals.

Female: The individual member of a species that produces eggs.

Fitness: Those qualities in an individual of a species that enable it to survive and reproduce in competition with other members of the same species.

Flehmen: A retraction of the upper lip which exposes the incisor teeth and gives access to the organ of Jacobson; performed by some male mammals after sniffing the urine of the female.

Gamete: A mature sex cell; the egg or sperm.

Gene: A single unit of heredity; a continuous length of DNA with a single genetic function.

Genetic fitness: The contribution to the next generation of one genotype relative to others in the population.

Genome: The complete genetic makeup of an individual.

Genotype: The genetic constitution of an individual with reference either to a single trait or a set of traits. (Contrast with *Phenotype.*)

Gonad: An organ that produces gametes; the ovary in the female and the testis in the male.

Habitat: The physical place where an organism lives and grows.

Haplodiploidy: A mode of sex determination during which males develop from unfertilized eggs and females from fertilized eggs.

Haploid: Having a single set of genes or chromosomes; eggs and sperm are haploid.

Harem: A group of females guarded by a male which prevents other males from mating with them. Occasionally, a group of males guarded by a female which prevents other females from mating with them.

Hermaphrodite: An organism which houses female and male sex organs, both of which are functional.

Herpetology: The scientific study of reptiles and amphibians.

Hormone: A chemical substance produced at a particular site in the body that travels through the blood stream to a target organ, where it exerts its influence.

Icthyology: The scientific study of fish.

Inclusive fitness: The sum of an individual's fitness plus all its influence on relatives other than offspring, where each effect is multipled by the actor's degree of relatedness to the affected party.

Infanticide: The killing of offspring by another member of its species, usually to provide resources for the other member's offspring or potential offspring.

Kin selection: The differential survival and reproduction of organisms owing to the behavior of relatives who possess some of the same genes through common ancestors.

Larva: An immature life form that differs in structure from the adult, such as a tadpole or caterpillar.

Laufschlag: A courting gesture made by some male ungulates which consists of touching the female's underparts with a stiff foreleg.

Lek: An arena used consistently for communal courtship displays.

Lordosis: A mating posture assumed by many female mammals in which the back is arched and the head and rump raised.

Male: In species that have two sexes, the individual that produces sperm.

Mammalogy: The scientific study of mammals.

Marsupial: A mammal belonging to the subclass Metatheria which gives birth to embryonic young that usually complete their development in a pouch.

Meiosis: The production of four haploid gametes through two successive divisions of a diploid cell.

Modern synthesis: The merging of the Darwinian idea of natural selection with Mendelian genetics, in which genes are explained as the hereditary units upon which the selective process acts.

Monogamy: The condition in which one male and one female join together to raise at least a single brood.

Monomorphic: Having a single body size, color, or form in both sexes.

Monotreme: A member of the mammalian subclass Prototheria which lays eggs. There are just two surviving members, the platypus and the spiny anteater (echinda).

Natural selection: The differential survival and reproduction of organisms with different genetic makeups. This is the basic evolutionary mechanism proposed by Charles Darwin and is still regarded as the major guiding force in evolution.

Niche: The totality of the adaptations, specializations, tolerance limits, functions, and biological interactions with other animals and plants of a species. Habitat is analogous to an animal's address and niche to its occupation.

Ornithology: The scientific study of birds.

Oviparous: Producing eggs that develop and hatch outside the mother's body.

Ovoviviparous: Producing young that develop within the mother's body but derive most or all of their nourishment from the egg yolk.

Parthenogenesis: Asexual reproduction in which eggs develop into organisms without having been fertilized.

Phenotype: The outward appearance of an organism. (Contrast with *Genotype.*)

Pheromone: A chemical substance used by one member of a species to communicate with another member of the same species.

Placenta: An organ formed in most female mammals to nourish the developing fetus and remove its waste. The placenta consists of both maternal and fetal tissues.

Polyandry: The acquisition by a female animal of several mates, either all at once or in sequence. It usually also follows that the males cooperate with the female in raising the young.

Polygyny: The acquisition by a male animal of several female mates, either all at once or in sequence.

Population: A group of animals of the same species occupying the same space or territory at the same time.

Precocial young: Animals that are born in an advanced state and are able to move and find food on their own at a very early age. (Contrast with *Altricial young.*)

Primate: The order of mammals which contains prosimians, monkeys, apes, and *Homo sapiens.*

Reciprocal altruism: The trading of altruistic acts by individuals.

Reproductive success: The number of surviving offspring of an individual or group of individuals.

Sexual dimorphism: Any consistent difference beyond the sex organs in the physical shape of females and males in the same species.

Sexual selection: Competition within one sex for access to the opposite sex and intersexual choice.

Society: A group of individuals of the same species which communicate and cooperate.

Species: A group of animals that naturally breed among themselves.

Spermatophore: A capsule or mass containing spermatozoa that is transferred by a male animal to a female.

Strategy: A maneuver aimed at taking advantage of an adversary or an environment.

Taxonomy: The science of classifying living organisms, usually according to their evolutionary relationships.

Territory: An area occupied by some species and defended for their own use or the use of their kin.

Trophallaxis: In social insects, the exchange of liquid from the alimentary system among members of the same colony.

Zygote: A fertilized egg which is created by the union of two gametes.

Bibliography

Adriaanse, M. S. C. 1947. *Behaviour* 1:1 – 35.

Altmann, Jeanne. 1974. "Observational study of behavior: sampling methods." *Behaviour* 49:227 – 267.

——— 1980. *Baboon Mothers and Infants.* Cambridge: Harvard University Press.

Anderson, Malte. 1982. "Female choice selects for extreme tail length in the widowbird." *Nature* 299:818 – 820 (October 28).

Arnold, Stevan J., and Paul H. Harvey. 1982. "Female mate choice and runaway sexual selection." *Nature* 297:533 – 534 (June 17).

Austin, C. R., and R. V. Short. 1972. *Reproduction in Mammals.* Vol. 4, *Reproductive Patterns.* Cambridge: Cambridge University Press.

Ayala, Francisco J. 1978. "The mechanisms of evolution." *Scientific American* 239(3):56 – 69 (September).

Baerends, G. P., and J. M. Baerends-Van Roon. 1950. "An introduction to the study of the ethology of cichlid fishes." *Behaviour* 1:1 – 242 (suppl.).

Bailey, Robert O., Seymour Norman, and Garey R. Stewart. 1978. "Rape behavior in blue-winged teal." *Auk* 95:188 – 190.

Barash, David. 1977. "Rape among mallards." *Science* 197:788 (August 19).

Barlow, George W. 1970. "A test of appeasement and arousal hypotheses of courtship behavior in a cichlid fish, *Etroplus maculatus.*" *Zeitschrift fur Tierpsychologie* 27:779 – 806.

Barlow, R. B., Jr., L. C. Ireland, and L. Kass. 1982. "Vision has a role in limulus mating behaviour." *Nature* 296:65 – 66 (March 4).

Bastock, Margaret. 1967. *Courtship: An Ethological Study.* Chicago: Aldine.

Bateson, Patrick, ed. 1983. *Mate Choice.* Cambridge: Cambridge University Press.

Beatty, Bill. 1979. *Unique to Australia.* Sydney: Paul Hamlyn.

Beer, C. G. 1983. "Darwin, instinct and ethology." *Journal of the History of the Behavioral Sciences* 19:68–79 (January).

Berger, Joel. 1983. "Induced abortion and social factors in wild horses." *Nature* 303:59–61 (May 5).

Bertram, Brian. 1975. "Social factors influencing reproduction in wild lions." *Journal of Zoology, London* 177:463–482.

——— 1975. "The social system of lions." *Scientific American* 232(5):54–65 (May).

——— 1979. "Ostriches recognise their own eggs and discard others." *Nature* 279:233–234 (May 17).

Blackwell, Antoinette Brown. 1875. *The Sexes throughout Nature.* New York: Putnam.

Bradbury, J. W. 1977. "Social organization and communication." In W. Wimsatt, ed., *The Biology of Bats,* vol. 3. New York: Academic Press, pp. 1–64.

——— 1983. "Leks and mate choice." In P. Bateson, ed., *Mate Choice.* Cambridge: Cambridge University Press, pp. 109–138.

Bradbury, J. W., and Robert M. Gibson. 1977. "Lek mating behavior in the hammer-headed bat." *Zeitschrift fur Tierpsychologie* 45:225–255.

Bristowe, W. S. 1958. *The World of Spiders.* London: Collins Press.

Brockleman, W. V., and D. Schilling. 1984. "Inheritance of stereotyped gibbon calls." *Nature* 312:634–636 (December 13).

Brown, L. Barton, ed. 1974. *Experimental Analysis of Insect Behavior.* New York: Springer-Verlag.

Brown, Margaret E. 1957. *The Physiology of Fishes.* Vol. 2, *Behavior.* New York: Academic Press.

Bruce, Hilda M. 1966. "Smell as an exteroreceptive factor." *Journal of Animal Science* 34:278–294 (suppl.).

Buchsbaum, Ralph, and Lorus J. Milne. 1967. *The Lower Animals: Living Invertebrates of the World.* New York: Doubleday.

Buechner, H. K., J. A. Morrison, and W. Leuthold. 1966. "Reproduction in Uganda kob, with species reference to behavior." In I. W. Rowlands, ed., *Comparative Biology of Reproduction in Mammals.* London: Academic Press, pp. 71–87.

Buechner, H. K., and H. D. Roth. 1974. "The lek system in Uganda kob." *American Zoologist* 14:145–162.

Buechner, H. K., and R. Schloeth. 1965. "Ceremonial mating behavior in Uganda kob." *Zeitschrift fur Tierpsychologie* 22:209–225.

Bull, J. J., R. C. Vogt, and J. McCoy. 1982. "Sex determining temperature in turtles: a geographic comparison." *Evolution* 36(2):326–332.

Burkhardt, Richard W. 1983. "The development of an evolutionary ethology." In D. S. Bendall, ed., *Evolution from Molecules to Men*. Cambridge: Cambridge University Press, pp. 429–444.

Burley, Nancy. 1982. "Facultative sex-ratio manipulation." *American Naturalist* 120:81–107.

Bursell, E. 1970. *An Introduction to Insect Physiology*. London: Academic Press.

Buskirk, Ruth E. 1981. "Sociality in the Arachnida." In Henry Rittermann, ed., *Social Insects*, vol. 2. New York: Academic Press, pp. 281–353.

Calder, William A. 1967. "Breeding behavior of the roadrunner *Geacaccyx californianus*." *Auk* 84:597–598.

Campbell, Bernard, ed. 1973. *Sexual Selection and the Descent of Man, 1871–1971*. Chicago: Aldine.

Campbell, Elisa K. 1983. "Beyond anthropocentrism." *Journal of the History of the Behavioral Sciences* 19:54–67 (January).

Carr, Archie. 1965. "The navigation of the green turtle." *Scientific American* 212(5):79–86 (May).

Chapman, R. F. 1982. *The Insects: Structure and Function*. 3rd ed. Cambridge: Harvard University Press.

Chism, Janice, Thelma Rowell, and Dana Olson. 1984. "Life history patterns of female patas monkeys." In Meredith F. Small, ed., *Female Primates: Studies by Women Primatologists*. New York: Alan R. Liss, pp. 175–192.

Clarke, Margaret P., and Kenneth E. Glander. 1984. "Female reproductive success in a group of free ranging howler monkeys *(Aloutta pallista)* in Costa Rica." In Meredith F. Small, ed., *Female Primates: Studies by Women Primatologists*. New York: Alan R. Liss, pp. 111–126.

Clutton-Brock, T. H. 1982. "Sons and daughters." *Nature* 298:11–13 (July 1).

——— 1982. "The red deer of Rhum." *Natural History* 91(11):43–46 (November).

Clutton-Brock, T. H., S. D. Albon, and F. E. Guiness. 1982. "Competition between female relatives in a matrilocal mammal." *Nature* 300:487–489 (November 11).

Clutton-Brock, T. H., and Paul Harvey. 1978. *Readings in Sociobiology*. San Francisco: W. H. Freeman.

Collias, Nicholas E. 1979. "Nest and mate selection by the female village weaver bird." *Animal Behaviour* 27:310.

Conover, Michael R., Don E. Miller, and George L. Hunter, Jr. 1979. "Female–female pairs and other unusual reproductive associations in ring billed and california gulls. *Auk* 96:6–9.

Cox, Cathleen, and Burney Le Boeuf. 1977. "Female incitation of male competition: a mechanism in sexual selection." *American Naturalist* 111:317–335.

Crews, David. 1975. "Psychobiology of reptilian reproduction." *Science* 189:1059–1065 (September 26).

——— 1979. "The hormonal control of behavior in a lizard." *Scientific American* 241(2):180–187 (August).

Crews, David, and K. Fitzgerald. 1980. "Sexual behavior in parthenogenetic lizards *(Cnemidophorus).*" *Proceedings of the National Academy of Sciences, U.S.A.* 77(1):499–502.

Crews, David, and William Garstka. 1982. "The ecological physiology of the garter snake." *Scientific American* 247(5):158–168 (May).

Crews, David, and Neil Greenberg. 1981. "Function and causation of social signals in lizards." *American Zoologist* 21:273–294.

Daly, Martin, and Margo Wilson. 1983. *Sex, Evolution and Behavior.* Boston: Willard Grant Press.

Darevsky, I. S. 1958. "Natural parthenogens in certain subspecies of rocky lizard, *Lacerta saxicola* Eversmann." *Dokl. Biol. Sci. Sect.* 122:877–879.

Darling, F. F. 1938. *Bird Flocks and the Breeding Cycle: A Contribution to the Study of Avian Sociality.* Cambridge: Cambridge University Press.

Darwin, Charles. 1859. *On the Origin of Species by Means of Natural Selection, or the Preservation of Favored Races in the Struggle for Life.* London: Murray.

——— 1871. *The Descent of Man and Selection in Relation to Sex.* London: Murray.

——— 1972. *The Expression of the Emotions in Man and Animals.* London: Murray.

Dawkins, R., and T. R. Carlisle. 1976. "Parental investment, mate desertion and a fallacy." *Nature* 262:131–133 (July 8).

Devine, Michael. 1975. "Copulatory plugs in snakes: enforced chastity." *Science* 187:844–845 (September 9).

De Vore, Irven. 1963. "Mother–infant relations in free ranging baboons." In Harriet B. Reingold, ed., *Maternal Behavior in Mammals.* New York: John Wiley and Sons, pp. 305–335.

Diamond, Jared. 1982. "Evolution of bowerbird's bowers: animal origins of the aesthetic sense." *Nature* 297:99–102 (May 13).

Dolhinow, Phyllis, and Nancy Krusko. 1984. "Langur monkey females and infants: the female's point of view. In Meredith F. Small, ed., *Female Primates: Studies by Female Primatologists.* New York: Alan R. Liss, pp. 35–57.

Douglas-Hamilton, Ian, and Ora Douglas-Hamilton. 1975. *Among the Elephants.* New York: Viking Press.

Drickhamer, L. C. 1974. "A ten year summary of reproductive data for free ranging *Macaca mulatta." Folia Primatologica* 21:61–68.

Dublin, Holly T. 1983. "Cooperation and reproductive competition among female African elephants." In Samuel K. Wasser, ed., *Social Behavior of Female Vertebrates.* New York: Academic Press, pp. 291–310.

Du Chaillu, Paul. 1861. *Explorations and Adventures in Equatorial Africa.* London: John Murray.

Eaton, R. L. 1969. "The social life of the cheetah." *Animals* 12(4):172–175.

Eberhard, William. 1985. *Sexual Selection and Animal Genitalia.* Cambridge: Harvard University Press.

Editorial. 1964. *Bee World* 45:133.

Ehrman, Lee. 1973. "Genetics and sexual selection." In B. Campbell, ed., *Sexual Selection and the Descent of Man,* 1871–1971. Chicago: Aldine, pp. 87–104.

Emlen, Stephen T. 1978. "The evolution of cooperative breeding in birds." In J. R. Krebs and N. B. Davis, eds., *Behavioral Ecology: An Evolutionary Approach.* Sunderland, Mass.: Sinauer Associates, pp. 245–282.

Emlen, S. T., and L. W. Oring. 1977. "Ecology, sexual selection and the evolution of mating systems." *Science* 197:215–223 (July 15).

Estes, Richard. In Press. *A Behavioral Guide to African Mammals.*

Ewer, R. F. 1973. "The evolution of mating systems in the Felidae." In R. Eaton, ed., *The World's Cats.* Winston, Ore.: World Wildlife Safari and Institute for the Study and Conservation of Endangered Species, pp. 110–160.

Foelix, R. F. 1982. *Biology of Spiders.* Cambridge: Harvard University Press.

Fossey, Dian. 1983. *Gorillas in the Mist.* Boston: Houghton Mifflin.

Frame, Lory Herbison, and George W. Frame. 1976. "Female African wild dogs emigrate." *Nature* 263:227–229 (September 16).

Franks, Nigel R., and Edward Scovell. 1983. "Dominance and reproductive success among slave-making ants." *Nature* 304:724–726 (August 25).

Frazer, J. F. D. 1959. *The Sexual Cycle of Vertebrates.* London: Hutchinson University Library.

Fricke, Hans W. 1979. "Mating system, resource defense and sex change in the anemonefish *Amphiprion akallopisos." Zeitschrift fur Tierpsychologie* 50:313–326.

Frith, H. J., and J. H. Calaby. 1969. *Kangaroos.* Melbourne: Canberra Press.

Fuyuma, Yoshiaki. 1984. "Gynogenesis in *Drosophila Melanogaster." Japanese Journal of Genetics* 59:91–96.

Galdikas, Biruté M. F. 1971. "Orangutan reproduction in the wild." In Charles E. Graham, ed., *Reproductive Biology of the Great Apes.* New York: Academic Press, pp. 281–300.

Garstka, William, and David Crews. 1981. "The role of the female in the initiation of the reproductive cycle of the snake *Thomnophis melanogaster.*" *American Zoologist* 21:960.

——— 1982. "Female control of male reproductive function in a Mexican snake." *Science* 217:1159–1160 (September 17).

Geddes, Patrick, and J. Arthur Thomson. 1901. *The Evolution of Sex.* London: Walter Scott.

Geist, V. 1968. "On the interrelation of external appearance, social behaviour and social structure of mountain sheep." *Zcitschrift fur Tierpsychologie* 25:199–215.

——— 1971. *Mountain Sheep: A Study in Behavior and Evolution.* Chicago: University of Chicago Press.

Gertsch, Willis J. 1949. *American Spiders.* Princeton: D. van Nostrand.

Ghiselin, Michael T. 1969. "The evolution of hermaphroditism among animals." *Quarterly Review of Biology* 44:189–208.

Giese, Arthur C., and John Pearse, eds. 1974–1977. *Reproduction of Marine Invertebrates.* 5 vols. New York: Academic Press.

Gilbert, Lawrence E. 1976. "Postmating female odor in *Heliconius* butterflies: a male contributed antiaphrodisiac." *Science* 193:419–420 (July 30).

Gilliard, E. Thomas. 1958. *Living Birds of the World.* New York: Doubleday.

——— 1963. "The evolution of bowerbirds." *Scientific American* 209(2):38–46 (August).

Goodall, Jane van Lawick. 1971. *In the Shadow of Man.* Boston: Houghton Mifflin.

Gould, Stephen J. 1980. "Death before birth, or a mite's *nunc dimittlis.*" In *The Panda's Thumb.* New York: W. W. Norton, pp. 69–75.

Gross, Mart R., and Richard Shine. 1981. "Parental care and mode of fertilization in ectothermic vertebrates." *Evolution* 35(4):775–793.

Grzimek, Bernhard. 1972. *The Animal Life Encyclopedia.* New York: Van Nostrand.

——— 1977. *Encyclopedia of Ethology.* New York: Van Nostrand Reinhold.

Gubernick, David J., and Peter H. Klopfer. 1981. *Parental Care in Mammals.* New York: Plenum Press.

Guest, W. C., and James L. Lasswell. 1978. "A note on courtship behavior and sound production in the red drum." *Copeia* 2:337–338.

Gustafson, Jill E., and David Crews. 1981. "Effect of group size and physiological state of a cagemate on reproduction in the parthenogenic lizard *Cnemidophorous uniparens.*" *Behavioral Ecology and Sociobiology* 8:267–272.

Gwynne, Darryl T. 1982. "Mate selection by female katydids." *Animal Behaviour* 30:734–738.
——— 1984. "Courtship feeding increases female reproductive success in bushcrickets." *Nature* 307:361–362 (January 26).

Hailman, Jack. 1978. "Rape among mallards." *Science* 201:280–281 (July 21).
Hamilton, W. D. 1963. "The evolution of altruistic behavior." *American Naturalist* 97:354–356.
Hamilton, William J. III, Curt Busse, and Kenneth S. Smith. 1982. "Adoption of infant orphan chacma baboons." *Animal Behaviour* 30:29–34.
Hanken, James, and Paul W. Sherman. 1981. "Multiple paternity in Belding's ground squirrel litters." *Science* 212:351–357 (April 17).
Hausfater, Glenn, Jeanne Altmann, and Stuart Altmann. 1982. "Long term consistency of dominance relations among female baboons *(papio cyanocephalus)*." *Science* 217:752–754 (August 20).
Heinrich, Bernd, and George A. Bartholomew. 1979. "The ecology of the African dung beatle." *Scientific American* 241(5):146–156 (November).
Herald, Earl S. 1961. *Living Fishes of the World.* New York: Doubleday.
Hildebrand, Milton. 1974. *Analysis of Vertebrate Structure.* New York: John Wiley and Sons.
Howard, R. D. 1978. "The evolution of mating systems in bull frogs, *Rana catesbiana*." *Evolution* 32:850–871.
Hrdy, Sarah Blaffer. 1976. "Care and exploitation of non-human primate infants by conspecifics other than the mother." In Jay Rosenblatt, Robert A. Hinde, Colin Beer and Marie-Claire Busnel, eds., *Advances in the Study of Behavior*, vol. 6. San Francisco: Academic Press, pp. 101–158.
——— 1977. *The Langurs of Abu.* Cambridge: Harvard University Press.
——— 1981. *The Woman That Never Evolved.* Cambridge: Harvard University Press.
——— 1981. "Nepotists' and 'altruists': the behavior of old females among macaques and langur monkeys." In Pamela T. Arnoas and Stevan Harrell, eds., *Other Ways of Growing Old.* Stanford: Stanford University Press, pp. 61–76.
Hrdy, Sarah Blaffer, and Daniel B. Hrdy. 1976. "Hierarchial relations among female hanuman langurs." *Science* 193:913–915 (September 3).
Hubbard, Ruth, M. Henifin, and B. Fried, eds. 1979. *Women Look At Biology Looking At Women: A Collection of Feminist Critiques.* Cambridge: Schenkman.
Huck, U. W. 1982. "Pregnancy block in laboratory mice as a function of social status." *Journal of Reproductive Fertility* 66:181–184.
Hunsaker, Don. 1977. *The Biology of Marsupials.* New York: Academic Press.

Hunt, George L., Jr., and Molly W. Hunt. 1977. "Female–female pairing in western gulls in southern California." *Science* 196:1466–1467 (June 24).

Hunt, G. L., Jr., A. L. Newman, M. H. Warner, J. C. Wingfield, and J. Kaiwi. 1984. "Comparative behavior of male–female and female–female pairs among western gulls prior to egg-laying." *Condor* 86:157–162.

Huxley, Julian, A. C. Hardy, and E. B. Ford, eds. 1954. *Evolution as a Process.* London: Allen and Unwin.

Innes, William T. 1966. *Exotic Aquarium Fishes.* New Jersey: Metaframe Corporation.

Itani, Junichirô. 1959. "Paternal care in the wild Japanese monkey, *Macaca fuscata fuscata.*" *Primates* 2(1):61–93.

Jacobson, Martin. 1965. *Insect Sex Attractants.* New York: Interscience.

——— 1972. *Insect Sex Pheromones.* New York: Academic Press.

Janetos, Anthony C. 1980. "Strategies of female male choice: a theoretical analysis." *Behavoiral Ecology and Sociobiology* 7:107–112.

Jarvis, J. U. M. 1981. "Eusociality in a mammal: cooperative breeding in naked mole-rat colonies." *Science* 212:571–573 (May 1).

Jenni, Donald A. 1974. "Evolution of polyandry in birds." *America Zoologist* 14:129–144.

——— 1979. "Female chauvinist birds." *New Scientist* 82:896–899.

Jenni, D. A., and G. Collier. 1972. "Polyandry in the American jacana." *Auk* 89:743–765.

Jolly, Alison. 1966. *Lemur Behavior: A Madagascar Field Study.* Chicago: University of Chicago Press.

——— 1985. *The Evolution of Primate Behavior.* New York: Macmillan.

Jones, Greta. 1980. *Social Darwinism and English thought: the interaction between biological and social theory.* New Jersey: Humanities Press.

Kasuya, Toshio, and Helene Marsh. 1984. "Life history and reproductive biology of the short-finned pilot whale, *Globiecephala macrorhychus,* off the Pacific coast of Japan." In William F. Perrin, Robert L. Brownell, Jr., and Douglas P. DeMaster, eds., *International Whaling Commission Proceedings of the Conference on Cetacean Reproduction: Estimating Parameters for Stock Assessment and Management.* Cambridge: International Whaling Commission.

Keenleyside, Miles H. A. 1979. *Diversity and Adaptation in Fish Behavior.* New York: Springer-Verlag.

Kessel, E. L. 1955. "The mating activities of balloon flies." *Systematic Zoology* 4(3):97–104.

Keverne, E. B., and C. de la Riva. 1982. "Pheromones in mice: reciprocal interaction between the nose and brain." *Nature* 296:148–150 (March 11).

Keverne, E. B., F. Levy, P. Poindron, and D. R. Lindsay. 1983. "Vaginal stimulation: an important determinant of maternal bonding in Sheep." *Science* 219:81–83 (January 7).

Kevles, Bettyann. 1980. *Thinking Gorillas.* New York: E. P. Dutton.

Kingsley, S. R. 1977. "Early mother-infant behavior in two species of great apes: *Gorilla gorilla gorilla* and *Pongo pygmaeus pygmaeus.*" *DODO: Journal of the Jersey Wildlife Preservation Trust* 14:55–65.

Kleiman, Devra. 1977. "Monogamy in animals." *Quarterly Review of Biology* 52:39–63.

Klopfer, Peter H., and Jack P. Hailman. 1967. *An Introduction to Animal Behavior: Ethology's First Century.* New Jersey: Prentice-Hall.

Koenig, Walter D., and Frank A. Pitelka. 1979. Relatedness and inbreeding avoidance: counterploys in the communally nesting acorn woodpecker. *Science* 206:1103–1105 (November 30).

Koenig, Walter D., Ronald L. Mumme, and Frank Pitelka. 1983. "Female roles in cooperatively breeding acorn woodpeckers." In Samuel K. Wasser, ed., *Social Behavior of Female Vertebrates.* New York: Academic Press.

Koetter, M. J. 1980. "Darwin, Wallace and the Origin of Sexual Dimorphism." *Proceedings of the Americal Philosophical Society* 124(3):203.

Kruuk, Hans. 1972. *The Spotted Hyaena: A Study of Predation and Social Behavior.* Chicago: University of Chicago Press.

Kummer, H. 1967. "Tripartite relations in hamandryas baboons." In S. A. Altmann, ed., *Social Communication among Primates.* Chicago: University of Chicago Press, pp. 63–71.

Kuroda, Suehisa. 1980. "Social behavior of pygmy chimpanzees." *Primates* 21(2):181–197.

——— 1984. "Interaction over food among pygmy chimpanzees." In R. L. Susman, ed., *The Pygmy Chimpanzee.* New York: Plenum, pp. 301–324.

——— 1984. "Rocking gesture as communicative behavior in the wild pygmy chimpanzees in Wamba, Central Zaire." *Journal of Ethology* 137:127–137.

Kushlan, James A. 1973. "Observations on maternal behavior in the American alligator, *Alligator mississippiensis.*" *Herpetologica* 29:256–257.

Labov, Jay B. 1980. "Factors influencing infanticidal behavior in wild male house mice *(Mus musculus).*" *Behavoiral Ecology and Sociobiology* 6:297–303.

Lancaster, Jane B. 1971. "Play mothering: the relations between juvenile females and young infants among free ranging vervet monkeys." *Folia Primatologia* 15:161–182.

Langman, V. A. 1977. "Cow–calf relationships in giraffes." *Zeitschrift fur Tierpsychologie* 43:264–286.

Le Boeuf, B. J., R. J. Whiting, and R. F. Gahtt. 1972. "Perinatal behavior of northern elephant seal females and their young." *Behaviour* 43:121–156.

Lederberg, Joshua. 1947. "Gene recombination and linked segregation in *Escherichia coli*." *Genetics* 32:505.

Lewontin, Richard C. 1978. "Adaptation." *Scientific American*. 239(3):212–230 (September).

Li, Stacy K., and Donald H. Owings. 1978. "Sexual selection in the three-spined stickleback." *Zeitschrift fur Tierpsychologie* 46:359–371.

Liem, Karel F. 1968. "Geographical and taxonomic variation in the pattern of natural sex reversal in the teleost fish order *Synbranchiformes*." *Journal of Zoology, London* 156:225–238.

Lloyd, J. E. 1966. *Studies on the Flash Communication System in Photinus Fireflies.* Miscellaneous Publications from the Museum of Zoology, University of Michigan, Ann Arbor, no. 130.

Lorenz, Konrad Z. 1970. *Studies in Animal Behavior.* Trans. R. D. Martin. Cambridge: Harvard University Press.

Luft, Joan, and Jeanne Altmann. 1982. "Mother baboon." *Natural History* 91(9):31–38 (September).

MacDonald, D. W. 1979. "Helpers in fox society." *Nature* 282:69–71 (November 1).

Mallory, Frank F., and Ronald J. Brooks. 1978. "Infanticide and other reproductive strategies in the collared lemming." *Nature* 273:144–146 (May 19).

Manning, A. 1967. *An Introduction to Animal Behavior.* Reading, Mass.: Addison-Wesley.

Marshall, Joe T., and Elsie R. Marshall. 1976. "Gibbons and their territorial songs." *Science* 193:235–237 (July 16).

Martin, R. D. 1982. "Et tu tree shrew." *Natural History* 91(8):26–33 (August).

Maslin, T. Paul. 1971. "Parthenogenesis in reptiles." *American Zoologist* 11:361–380.

Matthews, L. Harrison. 1978. *The Natural History of the Whale.* New York: Columbia University Press.

Matthews, Robert, and Janice R. Matthews. 1978. *Insect Behavior.* New York: John Wiley and Sons.

May, Robert. 1978. "The evolution of ecological systems." *Scientific American* 239(3):160 – 175 (September).

Maynard Smith, John. 1978. *The Evolution of Sex.* Cambridge: Cambridge University Press.

——— 1978. "The evolution of behavior." *Scientific American* 239(3):176 – 192 (September).

Maynard Smith, J., and M. G. Ridpath. 1972. "Wife sharing in the Tasmanian native hen *Tribonxy martierii:* a case of kin selection?" *American Naturalist* 106:447 – 452.

Mayr, Ernst. 1970. *Populations, Species and Evolution.* Cambridge: Harvard University Press.

——— 1972. "Sexual selection and natural selection." In Bernard Campbell, ed., *Sexual Selection and the Descent of Man.* Chicago: Aldine, pp. 87 – 104.

——— 1978. "Evolution." *Scientific American* 239(3):46 – 55 (September).

McCann, T. S. 1982. "Aggressive and maternal activities of female southern elephant seals." *Animal Behaviour* 30:268 – 276.

McCracken, Gary F. 1984. "Communal nursing in Mexican free-tailed bat maternity colonies." *Science* 223:1090 – 1091 (March 9).

McLean, Ian G. 1982. "The association of female kin in the Arctic ground squirrel *Spermophilus parryii.*" *Behavioral Ecology and Sociobiology* 10:91 – 99.

Mech, L. David. 1981. *The Wolf: The Ecology and Behavior of an Endangered Species.* Minneapolis: University of Minnesota Press.

Merrett, P., ed. 1978. *Arachnology: Seventh Annual Congress of the Zoological Society of London.* London: Academic Press.

Miller, David B. 1979. "The acoustic basis of mate recognition by female zebra finches." *Animal Behaviour* 27:376 – 380.

Milne, Lorus, Margery Milne, and Franklin Russell. 1975. *The Secret Life of Animals.* New York: E. P. Dutton.

Moehlman, Patricia D. 1979. "Jackal helpers and pup survival." *Nature* 277:382 – 383 (February 1).

Moore, Jim. 1984. "Female transfer in primates." *International Journal of Primatology* 5:537 – 589.

Moore, J., and Rauf Ali. 1984. "Are dispersal and inbreeding avoidance related?" *Animal Behaviour* 32:94 – 102.

Mosedale, Susan S. 1978. "Science corrupted: Victorian biologists consider 'the woman question.'" *Journal of the History of Biology* 11(1):1 – 55.

Moss, Cynthia. 1982. *Portraits in the Wild: Animal Behavior in East Africa.* Chicago: University of Chicago Press.

Mumme, Ronald L., Walter D. Koenig, and Frank Pitelka. 1983. "Reproductive competition in the communal acorn woodpecker: sisters destroy each other's eggs." *Nature* 306:583–584 (December 8).

Nordenskiöld, Erik. 1935. *The History of Biology: A Survey.* Trans. Leonard Bucknall Eyre. New York: Tudor.

Novick, Alvin. 1969. *The World of Bats.* New York: Holt, Rinehart, & Winston.

O'Brien, S. J., M. E. Roelke, L. Marker, A. Newman, C. A. Winkler, D. Meltzer, L. Colly, J. F. Evermann, M. Bush, and D. E. Wildt. 1985. "Genetic basis for species vulnerability in the cheetah." *Science* 227:1428–1434 (March 22).

Packer, Craig, and Anne E. Pusey. 1982. "Cooperation and competition within coalitions of male lions: kin selection or game theory." *Nature* 296:740–743 (April 14).

Perrone, Michael, and Thomas Zaret. 1979. "Parental care patterns of fishes." *American Naturalist* 113:351–361.

Petrie, Marion. 1983. "Female moorhens compete for small, fat males." *Science* 220:413–414 (April 22).

Pietsch, Theodore. 1976. "Dimorphism, parasitism and sex: reproductive strategies among deep sea ceratoid angler fishes." *Copeia* 4:781–793.

Policansky, David. 1982. "Sex change in plants and animals." *Annual Review of Ecological Systems* 13:471–495.

Purchon, R. D. 1977. *The Biology of the Mollusca.* 2nd ed. Oxford: Pergamon Press.

Ralls, Katherine. 1976. "Mammals in which females are larger than males." *Quarterly Review of Biology* 51:245–276.

Rand, A. S. 1967. "The adaptive significance of territoriality in iguanid lizards." *Lizard Ecology: A Symposium.* Columbia: University of Missouri Press.

Richard, Alison F. 1985. *Primates in Nature.* New York: W. H. Freeman.

Richard, A. F., and S. R. Schulman. 1982. "Sociobiology: primate field studies." *Annual Review of Anthropology* 11:231–255.

Riddleford, L. M. 1974. "The role of hormones in the reproductive behavior of female wild silkmoths." In L. Barton Browne, ed., *Experimental Analysis of Insect Behavior.* New York: Springer-Verlag, pp. 278–283.

Ridley, Mark. 1978. "Paternal care." *Animal Behaviour* 26:904–993.

Riedman, Marianne L., and Burney LeBoeuf. 1982. "Mother–pup separation and adoption in northern elephant seals." *Behavioral Ecology and Sociobiology* 11:203–215.

Robel, Robert J., and Warren B. Ballard, Jr. 1974. "Lek, social organization and reproductive success in the greater prairie chicken." *American Zoologist* 14:121–128.

Robertson, Ross. 1972. "Social control of sex reversal in a coral-reef fish." *Science* 177:1007–1009 (September 15).

——— 1973. "Sex changes under the waves." *New Scientist* 58:538–540 (May 31).

Robinson, M. H., and Barbara Robinson. 1978. "The evolution of courtship systems in tropical araneid spiders." In P. Merrett, ed., *Arachnology: Seventh Annual Congress of the Zoological Society of London.* London: Academic Press.

Romanes, George John. 1910. *Darwin and after Darwin: An Exposition of the Darwinian Theory and a Discussion of Post-Darwinian Questions.* Chicago: Open Court.

Romer, A. S., and Thomas Parsons. 1977. *The Vertebrate Body.* Philadelphia: W. B. Saunders.

Roth, Louis, and Robert Barth. 1967. "Sense organs employed by cockroaches in mating behaviour." *Behaviour* 28:58–94.

Rovner, Jerome, and Peter Witte. 1982. *Spider Communication.* Princeton: Princeton University Press.

Rowell, T. E. 1966. "Hierarchy in the organization of a captive baboon group." *Animal Behaviour* 14:430–443.

——— 1968. "The effect of temporary separation from their group on the mother–infant relationship of baboons." *Folia Primatologia* 9:114–122.

Rowell, T. E., R. A. Hinde, and Y. Spencer-Booth. 1964. "'Aunt'–infant interaction in captive rhesus macaques." *Animal Behaviour* 12:219–226.

Rowlands, I. W. 1966. *Comparative Biology of Reproduction in Mammals.* Symposia of the Zoological Society of London, no. 15. London: Academic Press.

Russell, Jay. In Press. *Lepilemur.*

Rutowski, Ronald L. 1980. "Courtship solicitation by females of the checkered white butterfly *Pieris protodia.*" *Behavioral Ecology and Sociobiology* 7:113–117.

Schal, Coby, and William J. Bell. 1982. "Ecological correlates of paternal investments of urates in a tropical cockroach." *Science* 218:170–172 (October 8).

Schaller, George B. 1964. *The Year of the Gorilla.* Chicago: University of Chicago Press.

——— 1972. *The Serengeti Lion: A Study of Predator–Prey Relations.* Chicago: University of Chicago Press.

Schenkel, Rudolf, and Lotte Schenkel-Hulliger. 1969. *Ecology and Behavior of the Black Rhinoceros,* Diceros bicornis. Hamburg and Berlin: Paul Parey.

Schmidt, Karl P., and Robert F. Inger. 1957. *Living Reptiles of the World.* New York: Doubleday.

Scott, Linda M. 1984. "Reproductive behavior of adolescent female baboons *(Papio anubis)* in Kenya." In Meredith F. Small, ed., *Female Primates: Studies by Women Primatologists.* New York: Alan R. Liss, pp. 77–100.

Sebeok, Thomas, ed. 1977. *How Animals Communicate.* Bloomington: University of Indiana Press.

Sekulic, Ranka. 1976. "A case of adoption in the roan antelope." *Mammal News Communications* 3:235–238 (November).

Selander, R. K. 1972. "Sexual selection and dimorphism in birds." In Bernard Campbell, ed., *Sexual Selection and the Descent of Man, 1871–1971.* Chicago: Aldine.

Seyfarth, Robert M. 1976. "Social relationships among adult female baboons." *Animal Behaviour* 24:917–938.

Shaw, Evelyn, and Joan Darling. 1984. *Female Strategies.* New York: Walker.

Sherman, Paul W. 1977. "Nepotisism and the evolution of alarm calls." *Science* 197:1246–1253 (September 23).

——— 1981. "Kinship, demography, and Belding's ground squirrel nepotism." *Behavioral Ecology and Sociobiology* 8:251–259.

——— 1981. "Reproductive competition and infantacide in Belding's ground squirrels and other animals." In R. D. Alexander and D. W. Tinkle, eds., *Natural Selection and Social Behavior: Recent Research and New Theory.* New York: Chiron Press, pp. 311–331.

Silk, Joan B., and Robert Boyd. 1983. "Cooperation, competition and male choice in matrilineal macaque groups." In Samuel K. Wasser, ed., *Social Behavior of Female Vertebrates.* New York: Academic Press, pp. 315–347.

Simmonds, K. E. 1955. "Studies on great crested grebes." *Avicultural Magazine* 61(1):3–13; 61(2):93–102; 61(3):135–146; 61(4):181–201; 61(5):235–316; 61(6):294–316.

Small, Meredith F., ed. 1984. *Female Primates: Studies by Women Primatologists.* New York. Alan R. Liss.

Smuts, Barbara. 1982. "Special relationships between adult male and female olive baboons *(Papio/Anubis).*" Ph.D. diss., Stanford University.

Stacey, Peter B. 1982. "Female promiscuity and male reproductive success in social birds and mammals." *American Naturalist* 120:51–64.

Sullivan, Walter. 1982. "In shark womb: fetus cannibalizes rivals." *New York Times,* December 7, 1982, pp. 17, 20.

Taborsky, Michael, and Dominique Limberger. 1981. "Helpers in fish." *Behavioral Ecology and Sociobiology* 8:143–145.

Taub, D. 1980. "Female choice and mating strategies among wild barbary macaques." In Donald G. Lindburg, ed., *The Macaques: Studies in Ecology, Behavior, and Evolution.* New York: Van Nostrand-Reinhold, pp. 287–344.

Teale, Edwin W. 1949. *The Insect World of J. Henri Fabre.* New York: Harper.

Thorpe, W. H., and M. E. W. North. 1982. "Vocal imitation in the tropical bou-bou shrike *Laniarius aethiopicus major* as a means of establishing and maintaining social bonds." *Ibis* 108(3):432–435.

Tilson, Ronald, and Richard R. Tenaza. 1976. "Monogamy and duetting in an old world monkey." *Nature* 263:320–321 (September 23).

Tinbergen, Niko. 1951. *The Study of Instinct.* Oxford: Oxford University Press.

——— 1960. "The evolution of behavior in gulls." *Scientific American* 203(6):117–130 (December).

Trivers, R. L. 1972. "Parental investment and sexual selection." In T. H. Clutton-Brock and Paul Harvey, eds., *Readings in Sociobiology.* San Francisco: W. H. Freeman, pp. 136–179.

——— 1985. *Social Evolution.* Menlo Park: Benjamin Cummings.

Vehrencamp, Sandra L. 1977. "Relative fecundity and parental effort in communally nesting anis." *Science* 197:403–404 (July 22).

——— 1982. "Body temperatures of incubating versus non-incubating roadrunners." *Condor* 84:203–207.

von Holst, D. 1974. "Social stress in the tree shrew: its causes and ethological consequences." In R. D. Martin, G. A. Doyle, and A. C. Walker, eds., *Prosimian Biology.* London: Duckworth Press, pp. 389–411.

von Schiller, Florian, and Maurice Dow. 1977. "Courtship behavior in *Drosophila:* sexual isolation or sexual selection?" *Zeitschrift fur Tierpsychologie* 43:304–310.

Waage, J. K. 1979. "Adaptive significance of postcopulatory guarding of mates and nonmates by male *Calopteryx maculata* Odonata." *Behavioral Ecology and Sociobiology* 6:147–154.

——— 1979. "Dual function of the damselfly penis: sperm removal and transfer." *Science* 203:916–918 (March 2).

Washburn, S. L., and Irven De Vore. 1961. "The social life of baboons." *Scientific American* 210(6):62–66 (June).

Wasser, Samuel K., ed. 1983. *Social Behavior of Female Vertebrates.* New York: Academic Press.

Whitfield, Philip, ed. 1979. *The Animal Family.* New York: Norton.

Wickler, Wolfgang. 1972. *The Sexual Code.* New York: Doubleday.

Wilbur, Karl, and C. M. Yonge, eds. 1966. *The Physiology of Mollusca.* New York: Academic Press.

Wilczynski, Jan Z. 1960. "On egg dimorphism and sex determination in *Bonellia viridis.*" *Journal of Experimental Zoology* 143:61–75.

Wiley, R. H. 1974. "Evolution of social organization and life history among grouse." *Quarterly Review of Biology* 49:201–227.

——— 1978. "The lek mating system of the sage grouse." *Scientific American* 238(5):114–125 (May).

Wilson, Edward O. 1972. *The Insect Societies.* Cambridge: Harvard University Press.

——— 1975. *Sociobiology: The New Synthesis.* Cambridge: Harvard University Press.

Wingfield, J. C., A. Newmar, G. L. Hunt, Jr., and D. J. Farner. 1980. "Androgen in high concentrations in the blood of female western gulls." *Naturwissenschaften* 67:514.

Wittenberger, James P. 1981. *Animal Social Behavior.* Boston: Duxbury Press.

Wolf, Larry. 1975. "'Prostitution' behavior in a tropical hummingbird." *Condor* 77:140–144.

Wolfe, Linda D. 1984. "Japanese macaque sexual behavior: a comparison of Arashiyama East and West." In Meredith F. Small, ed., *Female Primates: Studies by Women Primatologists.* New York: Alan R. Liss, pp. 141–158.

Wright, Patricia C. 1984. "Biparental care in *Aotus trivirgatus* and *Callicebus moloch.*" In Meredith F. Small, ed., *Female Primates: Studies by Women Primatologists.* New York: Alan R. Liss, pp. 59–73.

Yasukawa, K., and W. A. Searcy. 1982. "Aggression in female red-winged blackbirds: a strategy to ensure male parental investment." *Behavioral Ecology and Sociobiology* 11:13–17.

Zahavi, A. 1975. "Mate selection: a selection for a handicap." *Journal of Theoretical Biology* 53:205–214.

Zuckerman, Solly. 1932. *The Social Life of Monkeys and Apes.* New York: Harcourt Brace.

Zumpe, Doris, and R. P. Michael. 1968. "The clutching reaction and orgasms in the female rhesus monkey *(Macaca mulatta).*" *Journal of Endocrinology* 40:117–123.

Photograph Sources

page 5: Burney J. Le Boeuf

page 6: A. Cruickshank / Vireo

page 16: Timothy Clutton-Brock

page 22: Ron Garrison, 1985 / Zoological Society of San Diego

page 24: Suehisa S. Kuroda

page 36: Helena Fitch / Zoological Society of San Diego

page 47: Bob Barlow

page 51: Richard Estes

page 54: U. William Huck

page 62: Jonathan Mee / Steinhart Aquarium

page 68: Jonathan Mee / Steinhart Aquarium

page 78: Australian Information Service

page 87: Coby Schal / State University of New Jersey, Rutgers

page 88: Patrick McGinnis / Gombe Stream Research Center

page 93: Fran Allan, 1974 / Animals Animals

page 96: George Schaller

page 100: Jonathan Mee / Steinhart Aquarium

page 105: Ron Garrison, 1985 / Zoological Society of San Diego *page 110:* Patrick McGinnis / Gombe Stream Research Center

page 123: Harold F. Hirth / University of Utah

page 126: Raymond A. Mendez / Animals Animals

page 129: Ron Garrison, 1985 / Zoological Society of San Diego

page 130: Tom H. Logan / Florida Fish and Game Commission

page 132: Sea World, Inc.

page 135: Miriam Austerman

page 137: Miriam Austerman

page 139: H. Cruickshank / Vireo

page 143: Gail Rubin, 1974 / Photo Researchers

page 145: Zoological Society of San Diego

page 150: Miriam Austerman

page 152: Leonard Lee Rue, III / Animals Animals

page 155: Richard Estes

page 162: Miriam Austerman

page 171: Australian Information Service

page 176: John Morgan

page 178: Alfred M. Bailey / National Audubon Society

page 185: Biological Systems, Inc.

page 186: Gary F. McCracken / Department of Zoology, University of Tennessee

page 188: Paul Sherman

page 191: George D. Lepp / Bio-tec Images

page 199: Richard Estes

page 202: David Crews

page 204: Member of the research group with George Hunt / University of California at Irvine

page 206: Suehisa S. Kuroda

page 210: Miriam Austerman

page 214: Andrew Burke

page 218: Paul Sherman

page 222: Miriam Austerman

page 227: Miriam Austerman

Index